Evolutionary Ecology of Lizards

Evolutionary Ecology of Lizards

Editors

Francisco Javier Zamora-Camacho
Mar Comas

MDPI • Basel • Beijing • Wuhan • Barcelona • Belgrade • Manchester • Tokyo • Cluj • Tianjin

Editors
Francisco Javier Zamora-Camacho
Universidad Complutense de Madrid
Spain

Mar Comas
Universidad de Granada
Spain

Editorial Office
MDPI
St. Alban-Anlage 66
4052 Basel, Switzerland

This is a reprint of articles from the Special Issue published online in the open access journal *Diversity* (ISSN 1424-2818) (available at: https://www.mdpi.com/journal/diversity/special_issues/evolutionary_ecology_lizards).

For citation purposes, cite each article independently as indicated on the article page online and as indicated below:

LastName, A.A.; LastName, B.B.; LastName, C.C. Article Title. *Journal Name* **Year**, *Volume Number*, Page Range.

ISBN 978-3-0365-4051-1 (Hbk)
ISBN 978-3-0365-4052-8 (PDF)

Cover image courtesy of Mar Comas

Contents

About the Editors

Francisco Javier Zamora-Camacho (PhD) is a postdoctoral researcher at the National Museum of Natural Sciences (Spain). I am an evolutionary ecologist with a particular focus on amphibians and reptiles. Among others, I am particularly interested in ecotoxicology, ecoimmunology, predator–prey interactions, reproduction ecology, and the interrelationships among them. I obtained my PhD at the University of Granada (Spain), where I studied the interplay between thermoregulation, locomotion and the immune response in lizards along a 2200-m elevational gradient. As a postdoctoral researcher, I have studied the role of de-icing salt pollution as a trigger of evolution, and determinants of reproductive success in North-American frogs at Dartmouth College (USA). I have also worked on predator–prey interactions in American and Spanish amphibians, as well as the effects of nitrogenous compounds on natural and polluted-habitat amphibians, at the National Museum of Natural Sciences (Spain) and the Complutense University of Madrid (Spain). I have published 57 scientific articles, 51 of them in JCR journals. I am an author of four book chapters, and co-editor of two Special Issues.

Mar Comas (PhD) is a postdoctoral researcher at both the University of Granada (Spain) and at Dartmouth College (USA). My main interest relies in the evolution of life histories in a framework of interactions with parasites, altitudinal gradients, and local adaptation. I am particularly interested in the interplay among the immune system, oxidative stress, and telomere dynamics and its role in the evolution of senescence. I conducted research work in Morocco while studying for my Bachelor's degree at the University of Cadiz, with several publications and the description of a subspecies of Salamander. Then, I obtained a Master's degree in Biodiversity at the University of Barcelona (UB), learning skeletochronology and stable isotope analysis (SIA). I carried out my PhD at Doñana Biological Station **(EBD-CSIC)** with a Severo Ochoa contract. I learned genetic techniques for parasite identification and the study of the Major Histocompatibility Complex (MHC) at the Konrad Lorenz Institute for Ethology (Vienna, Austria). I have published 32 articles, 26 in JCR journals and 1 in an international journal of divulgation (FrogLog). I am author of two book chapters, editor of a book for CRC Press and co-editor of this Special Issue.

MDPI

Editorial

Evolutionary Ecology of Lizards: Lessons from a Special Issue

Francisco Javier Zamora-Camacho [1,*] and Mar Comas [2]

1 Departamento de Biodiversidad, Ecología y Evolución, Facultad de Ciencias Biológicas, Universidad Complutense de Madrid, C/José Antonio Novais 12, 28040 Madrid, Spain
2 Departamento de Zoología, Facultad de Ciencias, Universidad de Granada, 18071 Granada, Spain; marcomas@ugr.es
* Correspondence: zamcam@ugr.es

Citation: Zamora-Camacho, F.J.; Comas, M. Evolutionary Ecology of Lizards: Lessons from a Special Issue. *Diversity* **2021**, *13*, 565. https://doi.org/10.3390/d13110565

Received: 11 October 2021
Accepted: 12 October 2021
Published: 5 November 2021

Publisher's Note: MDPI stays neutral with regard to jurisdictional claims in published maps and institutional affiliations.

Regardless of taxonomical disquisitions on its yet unraveled phylogenetic relationships within and among taxa [1], the Lacertilia constitute one of the most successful clades within vertebrates. With above 6000 extant species, not only have lizards (as these animals are broadly referred to in common parlance) undergone an outstanding degree of diversification, but this comes along with adaptations that allow them to thrive in most terrestrial habitats worldwide and occupy astonishingly diverse ecological niches [2]. With the exception of Antarctica, lizards are represented in all continents, where they can be found from the seashore to alpine elevations, in a wide array of habitats encompassing deserts, temperate and tropical forests, meadowy, sandy or rocky grounds, and virtually all terrestrial landscapes available [2].

However, not all these habitats are equally relevant in lizard evolutionary history and speciation processes. Although reproductive isolation is deemed one of the most important motors of speciation, the reality is more intricate than that, and involves other variables limiting or potentiating gene flow [3]. Ultimately, reproductive isolation cannot simplistically be regarded as categorical: factors affecting the actual degree of gene flow impairment, the recurrence of isolation events, and their persistence through time determine the probability of speciation to culminate [4]. Therefore, habitats that are structured into fragmented landscapes, capable of acting as ecological islands, can be especially prone to promote radiation. That is the case of karstic landscapes, which constitute a mosaic of isle-like fragments composed of caves, hills, and towers where crevices abound and provide multitude of microhabitats that might boost speciation. This prediction was confirmed by Grismer et al. [5] in this Special Issue. In their work, they analyzed habitat preferences of 344 *Cyrtodactylus* gecko species (one the most diverse genera within vertebrates, distributed in Asia and Oceania), accounting for a phylogenetic reconstruction of the clade using mitochondrial DNA [5]. They concluded that not only is the preference for karstic landscapes the norm, but it also represents the ancestral condition, which confirms karsts as a major trigger of lizard diversity [5].

Notably, diversity of species with similar ecological requirements may result in conflicting forces for spatial distribution. According to contemporary niche theory, populational expansion and contraction processes interact with resource supply ratios and requirement overlaps to determine the probabilities of coexistence among species [6]. Rarely are these processes stagnant, as the establishment of biogeographical limits among species is usually the outcome of complex intra and interspecific dynamics that are not alien to phylogenetic relationships: more related species are more likely to overlap in their ecological requirements [7]. These premises were the foundation of the contribution by Escoriza and Amat [8] to this Special Issue. In their article, they categorized the niches occupied by the largest Lacertid lizards in Southwestern Europe according to climate and vegetation cover, controlled for a mitochondrial-DNA based phylogeny [8]. Their results underscore the key role of climate structuring species segregation, as well as the influence of vegetation cover maintaining partition within the overlapped areas [8]. They also confirmed that overlapping tends to be greater in closely-related species [8].

A way to escape overlapping could be specialization, be it by evolving disparate diets, by segregating activity periods, or by establishing in habitats where interspecific competitors are scant or even absent [9]. That could be the case of high-elevation specialists. While lizards at high elevations may enjoy competition release [10], they may also face harsh conditions such as reduced temperature, shrunk activity-time window, food scarcity, and low partial pressure of oxygen [11]. Surviving these extreme conditions could require a series of changes at the morphological, behavioral, or physiological level. If these changes are not plastic, the reversibility of the elevational ascension could be at stake. In their article published in this Special Issue, Gangloff et al. [12] tested whether *Iberolacerta bonnali* lizards, whose distribution range is confined to high elevations in the Pyrenees, have adjusted their physiology to low partial pressure of oxygen to such extent that a transplant to low elevation conditions would affect other aspects of their physiology. Gangloff et al. [12] detected that, compared with conspecifics maintained within their natural elevation range, translocated lizards select lower body temperatures in a controlled experiment, and that their locomotor performance at high body temperatures is impaired, which is consistent with a physiological adaptation to low partial pressure of oxygen making them obligatory high-elevation dwellers.

In other instances, however, the occupation of such extreme habitats as high elevations is facultative, which clears the way for comparative research lines: due to the aforementioned elevational shifts in environmental conditions, species that exist along elevational gradients provide exceptional scenarios for a wide array of evolutionary hypotheses [13]. Although special priority has been given to those particular traits that allow organisms to cope with the varying parameters described, other features whose potential link with elevation is less obvious have received little attention. That is the case of coloration. Most research on elevational variation in ectotherms' coloration focuses on melanization as a mechanism to improve heating in low-temperature environments, in light of the thermal melanism hypothesis [14]. Meanwhile, other color traits have been virtually neglected. In the article they contributed to this Special Issue, Moreno-Rueda et al. [15] advanced the filling of this research gap by studying various correlates of ornamental coloration of *Psammodromus algirus* lizards in the context of a 2200 m elevational gradient in Sierra Nevada mountains (SE Iberian Peninsula). Moreno-Rueda et al. [15] detected that the number of blue lateral eyespots decreases with increasing elevation, whereas throat color becomes more saturated. Along with the presence of a color patch in the mouth commissures, these traits indicate larger heads, while throat saturation and the occurrence of colored commissures are greater in older individuals [15]. Additionally, males have more eyespots, whereas the presence of a colored commissure is less frequent in females [15]. These findings suggest that different color patches may convey redundant information whose perceptibility can be tuned to different environmental or social circumstances. With elevation, both biotic and abiotic factors change, as well as the perceptibility of signals, with weather conditions, depending on altitude. Different perceptibility depending on light or weather conditions may strengthen the importance of the redundant information contained in different signals [15].

While lizards possess some traits, such as coloration, that are universal, or at least widespread, in the animal kingdom, others are, if not unique, restricted to a few taxa. One example is autotomy, which involves the capability of some animals to self-detach a body part to escape a predator's grasp [16]. Although the appendage released is usually not essential for life, autotomy oftentimes comes with a series of costs that could negatively redound upon the individual's fitness [17]. In the case of lizards, numerous species possess autotomic tails, whose loss is foreseeably accompanied by a reduction in locomotor performance [16,17]. However, this rule could not be universal. In this Special Issue, Silva et al. [18] published a work that tests whether caudal autotomy affects locomotion in the South American lizard *Micrablepharus atticolus*, a semifossorial species with reduced limbs that combines undulant movements with trot-like locomotion. They discovered that sprint speed was not affected by autotomy in this species, whereas it varied among collection sites, and was dependent on body temperature, body mass, reproductive status, and the length of the regenerated portion of the tail [18]. The latter results largely confirm well-known predictors of locomotor performance in animals. However, the lack of effect of tail autotomy on sprint speed suggests a role for the locomotion mode and the microhabitat used.

To conclude, lizards constitute exciting investigation subjects in evolutionary ecology for innumerable reasons (e.g., [19]). Facts as disparate as the complex relationships between the habitats they occupy and their diversification history, the ecological patterns of spatial segregation among species, the physiological strategies permitting their occurrence in extreme habitats, the intricate components of visual communication, and the potential consequences of extreme antipredator strategies have been explored in this Special Issue. In doing so, it has accomplished its crucial goal of bridging and forwarding the knowledge on the diverse disciplines of ecology to which the study of lizards represents an outstanding contribution.

Author Contributions: Conceptualization, F.J.Z.-C. and M.C.; writing-original draft preparation, F.J.Z.-C.; writing-review and edition, F.J.Z.-C. and M.C. All authors have read and agreed to the published version of the manuscript.

Conflicts of Interest: The authors declare no conflict of interest.

References

1. Ananjeva, N.B. Current state of the problems in the phylogeny of squamate reptiles (Squamata, Reptilia). *Biol. Bull. Rev.* **2019**, *9*, 119–128. [CrossRef]
2. Reilly, S.M.; McBrayer, L.B.; Miles, D.B. *Lizard Ecology*; Cambridge University Press: New York, NY, USA, 2007.
3. Barton, N.H. On the completion of speciation. *Philos. Trans. R. Soc. B* **2020**, *375*, 20190530. [CrossRef] [PubMed]
4. Harvey, M.G.; Singhal, S.; Rabosky, D.L. Beyond reproductive isolation: Demographic controls on the speciation process. *Ann. Rev. Ecol. Evol. Syst.* **2019**, *50*, 75–95. [CrossRef]
5. Grismer, L.; Wood, P.L.; Poyarkov, N.A.; Le, M.D.; Karunarathna, S.; Chomdej, S.; Suwannapoom, C.; Qi, S.; Liu, S.; Che, J.; et al. Karstic landscapes are foci of species diversity in the world's third-largest vertebrate genus Cyrtodactylus Gray, 1827 (Reptilia: Squamata; Gekkonidae). *Diversity* **2021**, *13*, 183. [CrossRef]
6. Letten, A.D.; Ke, P.J.; Fukami, T. Linking modern coexistence theory and contemporary niche theory. *Ecol. Monogr.* **2017**, *87*, 161–177. [CrossRef]
7. Godsoe, W.; Jankowski, J.; Holt, R.D.; Gravel, D. Integrating biogeography with contemporary niche theory. *Trends Ecol. Evol.* **2017**, *32*, 488–499. [CrossRef] [PubMed]
8. Escoriza, D.; Amat, F. Habitat partitioning and overlap by large lacertid lizards in southern Europe. *Diversity* **2021**, *13*, 155. [CrossRef]
9. Costa-Pereira, R.; Araújo, M.S.; Souza, F.L.; Ingram, T. Competition and resource breadth shape niche variation and overlap in multiple trophic dimensions. *Proc. R. Soc. B* **2019**, *286*, 20190369. [CrossRef] [PubMed]
10. Comas, M.; Escoriza, D.; Moreno-Rueda, G. Stable isotope analysis reveals variation in trophic niche depending on altitude in an endemic alpine gecko. *Basic Appl. Ecol.* **2014**, *15*, 362–369. [CrossRef]
11. Jacobsen, D. The dilemma of altitudinal shifts: Caught between high temperature and low oxygen. *Front. Ecol. Environ.* **2020**, *18*, 211–218. [CrossRef]
12. Gangloff, E.J.; Spears, S.; Kouyoumdjian, L.; Pettit, C.; Aubret, F. Does hyperoxia restrict Pyrenean rock lizards Iberolacerta bonnali to high elevations? *Diversity* **2021**, *13*, 200. [CrossRef]
13. Tito, R.; Vasconcelos, H.L.; Feeley, K.J. Mountain ecosystems as natural laboratories for climate change experiments. *Front. For. Glob. Chang.* **2020**, *3*, 38. [CrossRef]
14. Clusella Trullas, S.; van Wyk, J.H.; Spotila, J.R. Thermal melanism in ectotherms. *J. Therm. Biol.* **2007**, *32*, 235–245. [CrossRef]
15. Moreno-Rueda, G.; Reguera, S.; Zamora-Camacho, F.J.; Comas, M. Inter-individual differences in ornamental colouration in a Mediterranean lizard in relation to altitude, season, sex, age, and body traits. *Diversity* **2021**, *13*, 158. [CrossRef]
16. Emberts, Z.; Escalante, I.; Bateman, P.W. The ecology and evolution of autotomy. *Biol. Rev.* **2019**, *94*, 1881–1896. [CrossRef] [PubMed]
17. Maginnis, T.L. The costs of autotomy and regeneration in animals: A review and framework for future research. *Behav. Ecol.* **2006**, *17*, 857–872. [CrossRef]

18. Silva, N.A.; Caetano, G.H.O.; Campelo, P.H.; Cavalcante, V.H.G.L.; Godinho, L.B.; Miles, D.B.; Paulino, H.M.; da Silva, J.M.A.; de Souza, B.A.; Silva, H.B.F.; et al. Effects of caudal autotomy on the locomotor performance of Micrablepharus atticolus (Squamata, Gymnophthalmidae). *Diversity* **2021**, *13*, 562. [CrossRef]
19. Sinervo, B.; Livley, C.M. The rock-paper-scissors game and the evolution of alternative male strategies. *Nature* **1996**, *340*, 240–243. [CrossRef]

Article

Karstic Landscapes Are Foci of Species Diversity in the World's Third-Largest Vertebrate Genus *Cyrtodactylus* Gray, 1827 (Reptilia: Squamata; Gekkonidae)

Lee Grismer [1,*], Perry L. Wood, Jr. [2], Nikolay A. Poyarkov [3,4], Minh D. Le [5,6,7], Suranjan Karunarathna [8], Siriwadee Chomdej [9], Chatmongkon Suwannapoom [10], Shuo Qi [11], Shuo Liu [12], Jing Che [13], Evan S. H. Quah [1,14], Fred Kraus [15], Paul M. Oliver [16], Awal Riyanto [17], Olivier S. G. Pauwels [18] and Jesse L. Grismer [1]

1 Department of Biology, La Sierra University, 4500 Riverwalk Parkway, Riverside, CA 92505, USA; lgrismer@lasierra.edu (L.G.); evanquah@yahoo.com (E.S.H.Q.); jgrismer@lasierra.edu (J.L.G.)
2 Department of Biological Sciences and Museum of Natural History, Auburn University, Auburn, AL 36849, USA; perryleewoodjr@gmail.com
3 Department of Vertebrate Zoology, Biological Faculty, M. V. Lomonosov Moscow State University, 119234 Moscow, Russia; n.poyarkov@gmail.com
4 Joint Russian-Vietnamese Tropical Research and Technological Center, Nghia Do, Cau Giay, Hanoi 10000, Vietnam
5 Department of Environmental Ecology, Faculty of Environmental Sciences, University of Science, Vietnam National University, 334 Nguyen Trai Road, Hanoi 10000, Vietnam; le.duc.minh@hus.edu.vn
6 Central Institute for Natural Resources and Environmental Studies, Vietnam National University, 19 Le Thanh Tong Street, Hanoi 10000, Vietnam
7 Department of Herpetology, American Museum of Natural History, Central Park West at 79th Street, New York, NY 10024, USA
8 Nature Exploration and Education Team, B-1/G-6, Soysapura Housing Scheme, Moratuwa 10400, Sri Lanka; suranjan.karu@gmail.com
9 Research Center in Bioresources for Agriculture, Industry and Medicine, Department of Biology, Faculty of Science, Chiang Mai University, Chiang Mai 50200, Thailand; siriwadee@yahoo.com
10 School of Agriculture and Natural Resources, University of Phayao, Phayao 56000, Thailand; chatmongkonup@gmail.com
11 State Key Laboratory of Biocontrol/The Museum of Biology, School of Life Sciences, Sun Yat-sen University, Guangzhou 510275, China; qishuo1992@outlook.com
12 Kunming Natural History Museum of Zoology, Kunming Institute of Zoology, Chinese Academy of Sciences, 32 Jiaochang Donglu, Kunming 650223, China; liushuo@mail.kiz.ac.cn
13 Laboratory of Herpetological Diversity and Evolution, Kunming Institute of Zoology, Chinese Academy of Sciences, No.21, Qingsong Lu, Ciba, Panlong District, Kunming 650204, China; chej@mail.kiz.ac.cn
14 Lee Kong Chian Natural History Museum, National University of Singapore, 2 Conservatory Drive, Singapore 11737, Singapore
15 Department of Ecology and Evolutionary Biology, University of Michigan, Ann Arbor, MI 48109, USA; fkraus@umich.edu
16 Biodiversity and Geosciences Program, Queensland Museum, South Brisbane, QLD 4101, Australia; p.oliver@griffith.edu.au
17 Museum Zoologicum Bogoriense, Research Center for Biology, The Indonesian Institute of Sciences, Widyasatwaloka Building, Raya Jakarta Bogor, Km.46., Cibinong 16911, West Java, Indonesià; awal_lizards@yahoo.com
18 Royal Belgian Institute of Natural Sciences, Rue Vautier 29, B-1000 Brussels, Belgium; opauwels@naturalsciences.be
* Correspondence: lgrismer@lasierra.edu

Citation: Grismer, L.; Wood, P.L., Jr.; Poyarkov, N.A.; Le, M.D.; Karunarathna, S.; Chomdej, S.; Suwannapoom, C.; Qi, S.; Liu, S.; Che, J.; et al. Karstic Landscapes Are Foci of Species Diversity in the World's Third-Largest Vertebrate Genus *Cyrtodactylus* Gray, 1827 (Reptilia: Squamata; Gekkonidae). *Diversity* **2021**, *13*, 183. https://doi.org/10.3390/d13050183

Academic Editor: Eric Buffetaut

Received: 23 March 2021
Accepted: 23 April 2021
Published: 28 April 2021

Abstract: Karstic landscapes are immense reservoirs of biodiversity and range-restricted endemism. Nowhere is this more evident than in the world's third-largest vertebrate genus *Cyrtodactylus* (Gekkonidae) which contains well over 300 species. A stochastic character mapping analysis of 10 different habitat preferences across a phylogeny containing 344 described and undescribed species recovered a karst habitat preference occurring in 25.0% of the species, whereas that of the other eight specific habitat preferences occurred in only 0.2–11.0% of the species. The tenth category—general habitat preference—occurred in 38.7% of the species and was the ancestral habitat preference for *Cyrtodactylus* and the ultimate origin of all other habitat preferences. This study echoes the results of a previous study illustrating that karstic landscapes are generators of species diversity within *Cyrtodactylus* and not simply "imperiled arks of biodiversity" serving as refugia for relics. Unfortunately, the immense financial returns of mineral extraction to developing nations largely outweighs concerns for biodiversity conservation, leaving approximately 99% of karstic landscapes with no legal protection. This study continues to underscore the urgent need for their appropriate management and conservation. Additionally, this analysis supports the monophyly of the recently proposed 31 species groups and adds one additional species group.

Keywords: Indochina; Southeast Asia; phylogeny; Indo-Australian Archipelago; Bent-toed geckos; karst; conservation

1. Introduction

The dramatic topography of karstic landscapes composes some of the most surreal images of our world and has stirred the emotions of ancient artisans and natural historians for time on end. But not only are these crenulated, repeating layers of rugged terrain steeped in natural beauty (Figure 1), they are the only refuge for some of the most seriously endangered species on the planet [1]. Asia contains 8.35 million km^2 of karstic habitat with some of the most extensive concentrations ranging from China to western Melanesia (Figure 2). These formations are notable for their fragmented, island-like nature, with hills, caves, and towers forming archipelagos of habitat-islands stretching across broad geographic areas. This, and their fractured and eroded surfaces—which provide a myriad of microhabitats in which many taxonomic groups have specialized—have contributed to their extraordinarily high degrees of range-restricted endemism [2–5]. Karst formations are often referred to as "imperiled arks of biodiversity" [5]. However, a stochastic character mapping analysis of habitat preference using 243 species of the gekkonid genus *Cyrtodactylus*—the third most speciose vertebrate genus on the planet—indicated just the opposite [6]. Grismer et al. [6] demonstrated that karstic landscapes not only harbor range-restricted endemics, but have been the foci of speciation for the largest independent gekkonid radiations across all of Indochina and Southeast Asia. They went on to show that even in this ecologically labile genus, karst-associated species outnumbered by threefold all other species bearing other specific habitat associations. As such, this has transformed our view of karstic landscapes from that of "limestone museums" harboring relictual endemics, to platforms of speciation and generators of biodiversity across a broad *taxonomic* landscape (e.g., [7–10]).

Figure 1. The Dragon Back karst formation in Perlis State, Peninsular Malaysia.

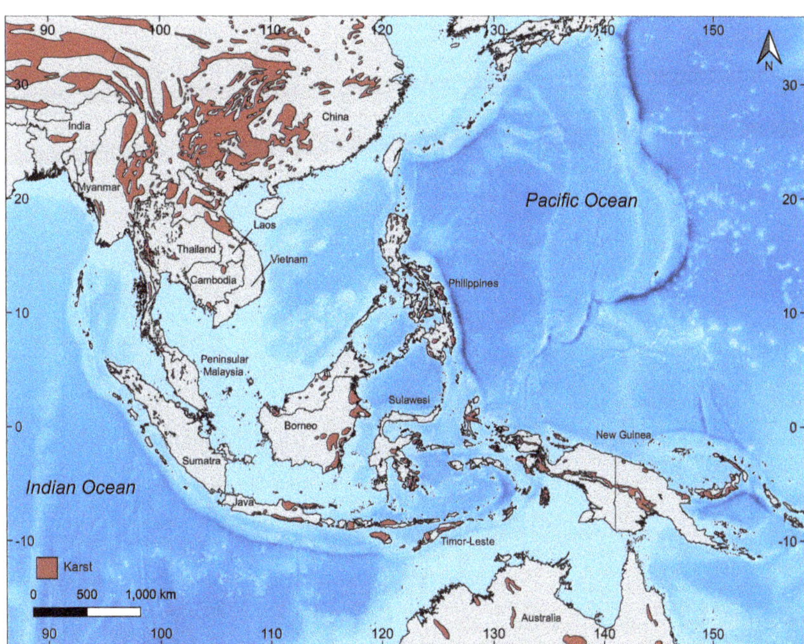

Figure 2. The distribution of karstic landscapes throughout Indochina and the Indo-Australian Archipelago.

Cyrtodactylus is by far the most speciose and ecologically diverse gekkotan genus [6,11]. It currently contains 306 nominal species (as of 14 February 2021; [12]) ranging from South Asia to Melanesia (Figure 3) where they occupy a vast diversity of habitats. As would be expected from a group this large and widely distributed, it bears a broad variety of ecotypes ranging from short robust terrestrial species to cryptically colored arboreal species to gracile cave-dwelling and karst-adapted specialists (e.g., [13–21]; Figure 4). The annual rate at which new species are being described is unprecedented and shows no signs of leveling off (Figure 5) and the majority of the most recently described species are associated with karst formations. In some cases, multiple species from distantly related clades may be found throughout a single karstic archipelago [14,20], and even more remarkable, different species from distantly related clades may even occupy the same small karst formation [14,20]. The intent of this paper is to test, (1) whether or not the same clades bearing the same specific habitat preferences presented by Grismer et al. [6] are recoverable, (2) whether or not the relative frequencies of species in each habitat preference category are not significantly different than that reported by Grismer et al. [6], (3) and specifically, is the hypothesis that karstic landscapes are generators of biodiversity further supported. We test these hypotheses by augmenting Grismer et al.'s [6] original phylogeny of 243 species with an additional 101 species (a 44% increase in species coverage) and by adding a new category of habitat preference. Additionally, with this significant influx of species, we test the monophyly of the 31 different species groups recently designated by Grismer et al. [11] based on their phylogeny of 310 named and unnamed species (an 11% increase in species coverage).

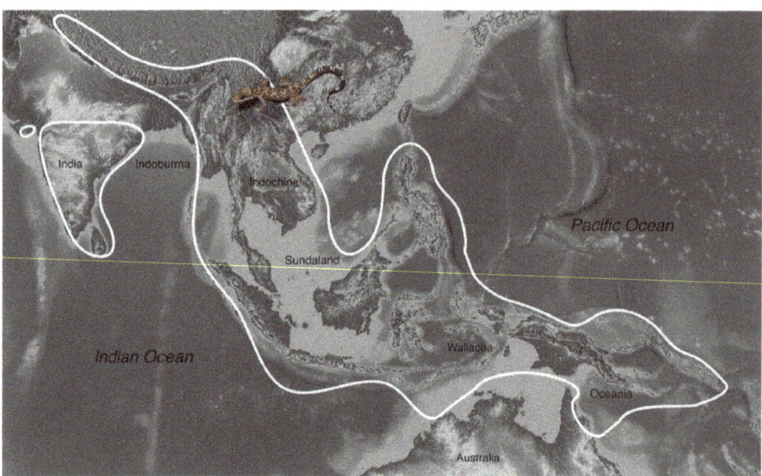

Figure 3. Generalized distribution of the genus *Cyrtodactylus*.

Figure 4. Representative ecotypes of the 10 different habitat preferences in the genus *Cyrtodactylus*. Photographs by (**A**) L. Lee Grismer, (**B**) Steve J. Richards, (**C**,**D**) L. Lee Grismer, (**E**) Suranjan Karunarathna, (**F**) L. Lee Grismer. (**G**) Evan S. H. Quah, (**H**,**I**) L. Lee Grismer, and (**J**) Peter Geissler.

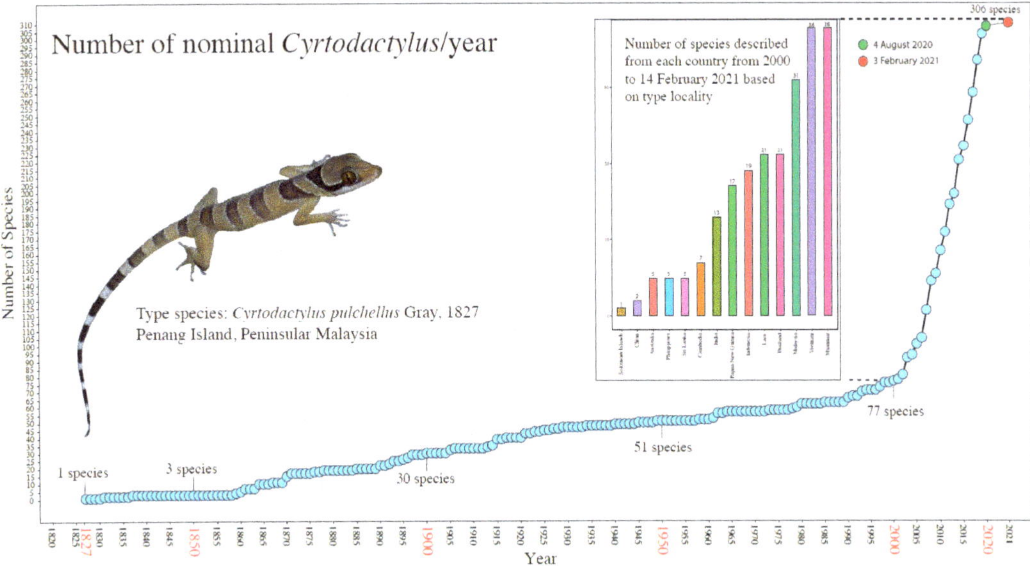

Figure 5. Cumulative number of species of *Cyrtodactylus* described per year. The trajectory of new species descriptions from 2000 to 18 March 2021 indicates that the true diversity of this genus is not yet calculable and that 48% of the newly described species during this period have come from Myanmar, Vietnam, and Malaysia.

2. Materials and Methods

2.1. Habitat Preferences and Ecotypes

Here we refine some of the criteria for designating habitat preference used by Grismer et al. [6] based on newly acquired data from recent publications and fieldwork. We also add an additional habitat preference (sandstone), bringing the total to 10 as opposed to nine categories (Table S1). Habitat preference for each species was coded as a discrete character state and ascertained by integrating data from the literature, firsthand experience of the authors, and personal communication with researchers familiar with particular species. Grismer et al. [6] acknowledged that some of these categories could be further subdivided (e.g., arboreal into branch, twig, and leaf), but those subdivisions become far less defensible owing to a lack of detailed microhabitat information. In this regard, many species can be considered data deficient, inasmuch as baseline information on their ecological requirements are often limited to anecdotal observations made at the time of their collection (e.g., [14]). The potential biases of using limited observations from a single locality at one point in time to ascertain the habitat preference of an entire species does not go unnoticed. However, in many cases, these are the only data available. Nonetheless, judiciously vetted, natural history observations summarized across the literature coupled with our own field observations and those of others, can provide a useful framework for supporting robust, testable, downstream hypotheses regarding habitat preference. The habitat preferences and their associated ecotypes bearing the same categorical names are described below. Obvious morphological correlates associated with some ecotypes are noted only for additional clarity.

1. General (Figure 4A). Species that use the majority of the microhabitats in their immediate surroundings in whatever environment they inhabit. The microhabitats may include rocks of all types (when present), logs, tree trunks (with or without holes and crevices), and all vegetative structures of various dimensions, the ground, and human-made structures in many cases. No particular microhabitat is notably preferred over any other although some species may be most often observed in low vegetation.

2. Trunk (Figure 4B). These are species generally found on the trunks and large branches of large trees at varying heights and often take refuge in cracks, crevices, or holes in the trunks. They may occasionally occur on large granite rocks but only if the rocks are near the trees. These species are generally the largest and most robust species in the genus [22–24]. None have been reported to have prehensile tails although some species may coil the tail horizontally similar to that seen in arboreal species.
3. Karst (Figure 4C). These are generally more gracile species that are restricted to habitats where limestone rock (karst) is present. Individuals use this substrate (including cliff faces, small rocks, and boulders) as well as adjacent vegetation. If caves are present, they will enter only into the twilight zone and usually no deeper than 50 m from the entrance [14]. Despite what has been written about many karst-associated species being cave species or cave adapted (e.g., [25]), none truly are and most are more commonly found on the outside of caves (see below). These species do not occur in habitats lacking karstic substrates.
4. Cave (Figure 4D). These are species that occur exclusively in the cave-like environments formed by large granite boulders. Open spaces between the boulders can be quite extensive and contain areas where very little light penetrates. These species rarely occur on the out-facing (i.e., the forest-side) surfaces of the boulders and for the most part, are restricted to the spaces between the boulders at varying depths below the surface of the ground in extremely low levels of illumination. These are truly cave-adapted species with notably thin, gracile bodies, long limbs, flat heads, large eyes, and faded color patterns [13,26,27].
5. Terrestrial (Figure 4E). These are species that generally occur only on the ground and may take refuge beneath natural and human-made surface objects. They may occasionally be found on the tops of small rocks (when present) or on the bases of small trees and shrubs but never higher than 1 m above the ground. These species are relatively small and notably squat, with short fat tails, thick heads, and short digits [28,29].
6. Arboreal (Figure 4F). These are cryptically colored species [30,31] generally restricted to small branches, leaves, trunks of varying sizes, and shrubs. Some may take refuge beneath exfoliating bark often as high or higher than three meters above the ground. These species are rarely observed on the ground or lower than 1.5 m above the ground. In such instances, it is usually during windy and/or rainy nights (perhaps forced down from higher up; [32]; authors pers. obs.) or during egg laying. All species have a prehensile tail used as a climbing aid [31–33] that is often carried in a coiled, elevated position.
7. Swamp (Figure 4G). These are species restricted to swampy habitats that use low, viny vegetation, the trunks of small trees and shrubs, or small logs often above, but always in close proximity to water. These species generally have large eyes with notably reddish-orange irises [34,35].
8. Granite (Figure 4H). These are generally more robust, strongly tuberculated species found in forested habitats bearing large granite boulders (not just small, scattered, granite rocks or rocks of other types). Vegetation is often used, especially by hatchlings and juveniles, but individuals occur more commonly on the granite boulders in all planes of orientation. These species do not occur in forested areas lacking granite boulders.
9. Intertidal (Figure 4I). This category contains a single species that occurs exclusively in the rocky intertidal zones of small islands in the Seribuat Archipelago off the southeastern coast of Peninsular Malaysia and avoids nearby forested regions even if they lack other species of *Cyrtodactylus* [19,36].
10. Sandstone (Figure 4J). This category was not included in Grismer et al. [6]. It contains a single species endemic to a forested sandstone massif isolated in the lowlands of northwestern Cambodia [11]. This species is known to forage only on the surface or within crevices of sandstone rocks and was not observed on the nearby vegetation [37].

This species is similar in body shape to closely related granite-associated species (Grismer unpublished).

2.2. Mitochondrial DNA

The data set of Grismer et al. [6] was augmented with 107 additional ingroup species resulting in a matrix composed of 344 described and undescribed species (i.e., species identified in previous phylogenies but not yet described) of *Cyrtodactylus* (Table S1). A phylogeny was constructed using 1474 base pairs of the mitochondrial gene NADH dehydrogenase subunit 2 gene and its flanking tRNAs (hereafter referred to as ND2). *Agamura persica, Bunopus tuberculatus, Hemidactylus angulatus, H. frenatus, H. garnotii, H. mabouia, H. turcicus, Lialis jicari, Mediodactylus russowii, Mokopiriakau cryptozoicus, Pygopus nigriceps, Sphaerodactylus torrei, Stenodactylus petrii, Tenuidactylus elongatus, Toropuku stephensi,* and *Tropiocolotes steudneri*—encompassing all other major gekkotan lineages—were used to root the tree following Wood et al. [38]. Genomic DNA was isolated from liver or skeletal muscle from new tissue samples stored in 95% ethanol, using standard phenol-chloroform-proteinase K (final concentration 1 mg/mL) extraction procedures with subsequent isopropanol precipitation following Hillis et al. [39] or a SPRI magnetic-bead extraction protocol (https://github.com/phyletica/lab-protocols/blob/master/extraction-spri.md; accessed on 15 January 2021). The ND2 gene, with parts of adjacent tRNAs, was amplified using a double-stranded Polymerase Chain Reaction (PCR) under the following conditions: 1.0 µL genomic DNA (10–30 µg), 1.0 µL light-strand primer (concentration 10 µM), 1.0 µL (H5934, 5′– AGRGTGCCAATGTCTTTGTGRTT–3′, following [6]), heavy-strand primer (concentration 10 µM), (L4437b, 5′–AAGCAGTTGGGCCCATRCC–3′, following [6]) 1.0 µL dinucleotide pairs (1.5 µM), 2.0 µL 5 buffer (1.5 µM), MgCl 10× buffer (1.5 µM), 0.1 µL Taq polymerase (5 u/µL), and 6.4 µL ultra-pure H_2O. PCR reactions were executed on Bio-Rad T100™ gradient thermocycler under the following conditions: initial denaturation at 95 °C for 2 min, followed by a second denaturation at 95 °C for 35 s, annealing at 55 °C for 35 s, followed by a cycle extension at 72 °C for 35 s, for 31 cycles. All PCR products were visualized using 1.0% agarose gel electrophoresis. Successful PCR products were sent to Evrogen® (Moscow, Russia), Genetech Sri Lanka Pvt. Ltd. (Colombo, Sri Lanka), or Genewiz® (South Plainfield, NJ, USA) for PCR purification, cycle sequencing, sequencing purification, and sequencing using the same primers as in the amplification step. Sequences were analyzed from both the 3′ and the 5′ ends separately to confirm congruence between reads. Forward and reverse sequences were uploaded and edited in GeneiousTM 2019.0.4 (https://www.geneious.com). Following sequence editing, the protein-coding region and the flanking tRNAs were aligned using the MAFTT v7.017 [40] plugin under the default settings in Geneious™ 2019.0.4 (https://www.geneious.com). Mesquite v3.04 [41] was used to calculate the correct amino-acid reading frame and to confirm the lack of premature stop codons in the ND2 portion of the DNA fragment.

2.3. Phylogenetic Analyses

A Maximum likelihood (ML) analysis was implemented using the IQ-TREE webserver [42,43] preceded by the selection of substitution models using the Bayesian Information Criterion (BIC) in ModelFinder [44] which selected TVM+F+I+G4 for the tRNAs and codon position 1 and GTR+F+I+G4 for codon positions 2 and 3. One-thousand bootstrap pseudoreplicates via the ultrafast bootstrap (UFB; [45]) approximation algorithm were employed, and nodes having UFB values of 95 and above were considered strongly supported [46]. We considered nodes with values of 90–94 as well supported.

A Bayesian inference (BI) phylogeny was estimated using Bayesian Evolutionary Analysis by Sampling Trees (BEAST) version 2.4.6 [47] implemented in CIPRES (Cyberinfrastructure for Phylogenetic Research; [48]). Input files were constructed in Bayesian Evolutionary Analysis Utility (BEAUti) version 2.4.6 using a lognormal relaxed clock with unlinked site models, linked trees and clock models, and a Yule prior and run in BEAST version 2.4.6 [47] on CIPRES. bModelTest, implemented in BEAST, was used to numerically integrate over the uncertainty of substitution models while simultaneously estimating phylogeny using Markov chain Monte Carlo (MCMC). MCMC chains were run for 350,000,000 generations and logged every 35,000 generations. The BEAST log file was visualized in Tracer v. 1.6.0 [49] to ensure effective sample sizes (ESS) were well above 200 for all parameters. A maximum clade credibility tree using mean heights at the nodes was generated using TreeAnnotator v. 1.8.0 [50] with a burn-in of 1000 trees (10%). Nodes with Bayesian posterior probabilities (BPP) of 0.95 and above were considered strongly supported [51,52]. We considered nodes with values of 0.90–0.94 as well supported.

Grismer et al. [6] demonstrated that in their 243-species data set, the third codon position contributed significantly to the strongly supported topological resolution of the tree and showed no signs of codon saturation. In their 310-species tree, Grismer et al. [11] demonstrated that their mito-nuclear tree constructed from ND2 and three nuclear genes did not improve the resolution or the nodal support of the deep nodes in their ND2 tree. Therefore, only ND2 was used in this analysis.

2.4. Ancestral State Reconstruction

In order to estimate the probability of each habitat preference at each node in the tree, we employed a stochastic character mapping (SCM) analysis implemented in R [v3.4.3] using the R package Phytools [53] on the BEAST tree converted to newick format. The transition-rate matrix that best fit the data was identified by comparing corrected Akaike Information Criterion (AICc) scores among alternate models using the R package ape 5.2 [54]. Three transition-rate models were considered: a 90-parameter model having different rates for every transition type (the ARD model); a 45-parameter model with equal forward and reverse rates between states (the symmetrical rates SYM model); and a single-rate parameter model that assumes equal rates among all transitions (ER). Lastly, an MCMC approach was used to sample the most probable 1000-character histories from the posterior using *make.simmap*() and then summarized them using the *summary*() command.

3. Results

The ML analysis recovered essentially the same well to strongly supported tree (Figure 6) recovered in Grismer et al. [11]. The same 31 monophyletic species groups designated in Grismer et al. [11] were recovered here even though sampling in was greatly expanded with additional species (Table S1). The ML analysis also recovered a new clade, designated here as the *tibetanus* group, that is composed of *Cyrtodactylus tibetanus*, *C.* cf. *tibetanus*, and *C. zhaoermii*. *Cyrtodacylus* cf. *tibetanus* and *C. zhaoermii* were unavailable for the analysis of Grismer et al. [11], where *C. tibetanus* was recovered as the earliest diverging member of the *lawderanus* group. Che et al. [55] recovered the same new clade in a less inclusive (i.e., fewer species) mito-nuclear phylogeny. Although Grismer et al. [11] recovered *C. rubidus* as the sister species of the *lateralis* group, it was not included in that group because this relationship was well supported only in the ML analysis and not the BI analysis. Here, it is placed in the *lateralis* group with high support in both analyses (90 UFB, 0.90 BPP), a grouping also supported by the fact that all members of this group have prehensile tails.

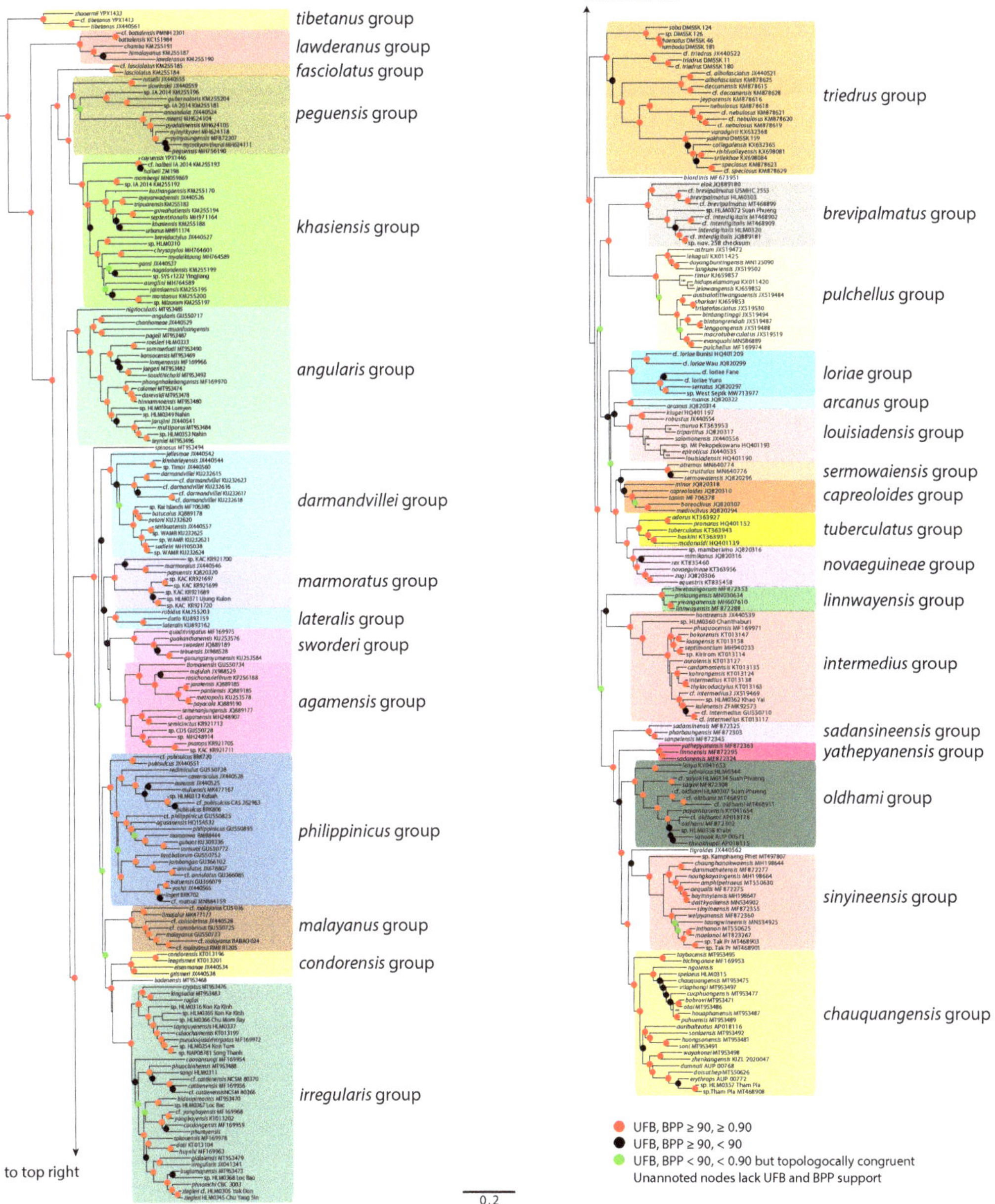

Figure 6. Majority-rule consensus tree from ML bootstrap replicates of 344 species of Cyrtodactylus. Phylogeny based on 1474 base pairs of the mitochondrial gene ND2 illustrating the designation of 32 monophyletic species groups.

The BEAST analysis recovered a tree with generally strong nodal support throughout with a 94.4% topological consistency (recovering 322 of the same 347 nodes) as the ML tree (Figure 7). The AICc scores for the three transition-rate models were ARD = 1101.751; SYM = 1035.445; and ER = 890.9552. The results of the SCM analysis were consistent with those of Grismer et al. [6] in that the ER model recovered large and small clades that independently evolved the same habitat preferences throughout the geographic range of the genus (Figure 7A). The SCM recovered a general habitat preference as being ancestral for not only the genus *Cyrtodactylus* but for all other major clades and ultimately all other habitat preferences as well. Notably for this study, however, the two largest independently evolved lineages of karst-associated species—the lineage composed of the *sadansinensis*, *yathepyanensis*, *oldhami*, *sinyineensis*, and *chauquangensis* groups and the *angularis* group—were also recovered, even with their expanded species contents. Their parapatric distributions across much of western and northern Indochina coincide with regions bearing the most extensive karstic landscapes (Figure 8). Other less diverse, independently evolved karstic lineages, such as the *linnwayensis* group from the Shan Plateau in Myanmar and a karst-associated subclade from the Thai-Malay Peninsula in the *pulchellus* group, were also recovered and associated with regions rich in karstic habitats (Figures 7A and 8). Several isolated instances of the independent evolution of karst habitat preference are scattered across the tips of the tree, representing species from Borneo (*C. cavernicolus*, *C. limajalur*, *C. muluensis*), Cambodia (*C. laangensis*), China (*Cyrtodactylus* sp. SYS r1232), Indonesia (*C. darmandvillei*), Myanmar (*C. aunglini*, *C. chrysopylos*, *C. myaleiktaung*), Papua New Guinea (*C. tanim*), Peninsular Malaysia (*C. evanquahi*, *C. guakanthanensis*, *C. gunungsenyumensis*, *C. metropolis*, *C. lenggongensis*, *C. sharkari*), and Vietnam (*C.* sp. nov., *C. yangbayensis*) (Figure 7A).

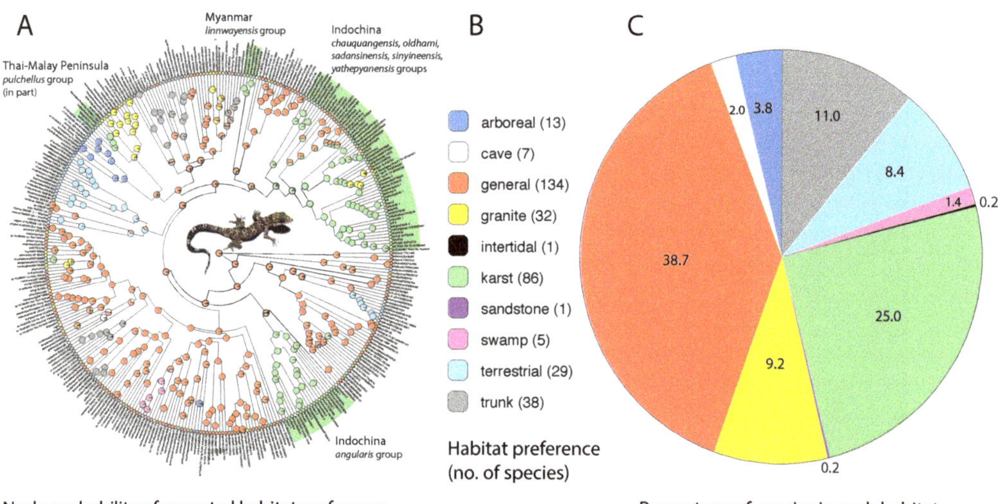

Figure 7. (**A**) Stochastic character map of the 10 habitat preferences on a maximum clade credibility BEAST phylogeny showing the probability of the ancestral habitat preference at each node and the major clades of karst-associated species groups and their general geographic distribution. (**B**) The habitat preference of each species, and the number of species in each habitat preference category. (**C**) Pie chart showing the percentage of species bearing each of the 10 habitat preferences.

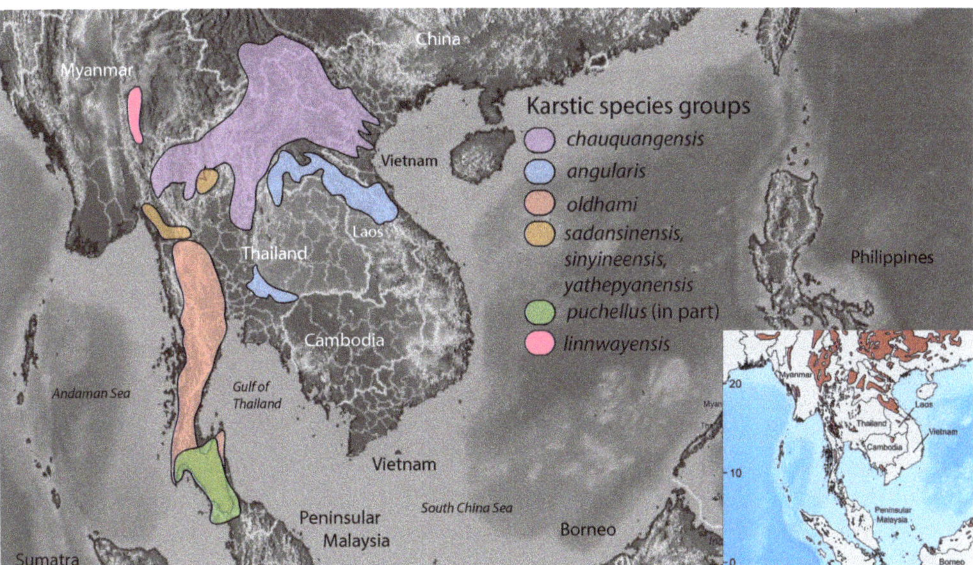

Figure 8. Distribution of the major clades of the karst-associated species groups throughout Indochina. Inset illustrates their co-distributions, with the geographic areas bearing the most extensive karstic landscapes.

These data are consistent with those of Grismer et al. [6] in showing that the frequency of karst-associated species far out-numbers that of any other specific habitat preference and is nearly two and one-half times more prevalent than any other specific habitat preference in that it contains 25.0% of the species followed by trunk (11.0%), granite (9.2%), terrestrial (8.4%), arboreal (3.8%), cave (2.0%), swamp (1.4%), and intertidal and sandstone (0.2%; Figure 7B). In Grismer et al. [6], granite-associated species comprised the second highest habitat preference and trunk-associated species the third. That ranking has been reversed here. The percentage of species with a karst habitat preference was 29.6% in Grismer et al. [6] but dropped to 25.0% here. We posit that this drop of nearly 5% is a direct result of our inability to explore unsurveyed karstic regions on the Shan Plateau and in the Salween Basin of Myanmar during 2020 due to COVID-19.

4. Discussion

The analysis presented here is based on the most complete phylogeny of the genus *Cyrtodactylus* to date with an increase of 101 species from that of Grismer et al. [6] and 35 from that of Grismer et al. [11]. The hypotheses marshaled by Grismer et al. [6] concerning the evolution of habitat preference is supported here in that there was no notable change in the frequencies of species bearing different habitat preferences across the genus—even with the addition of 107 species. More specifically, however, a karst habitat preference retained a higher frequency than that of any other specific habitat preference (25.0% versus 0.2–11%), supporting the hypothesis that these landscapes are platforms for the generation of biodiversity. This pattern is particularly strong in Indochina and less so on islands throughout the Indo-Australian Archipelago, reflecting the sharp contrast in the extent of karstic landscapes between these regions (Figure 8). These data clearly underscore the importance of karstic habitats to this hyper-diverse genus and continue to amplify the work of many other authors indicating that the high levels of biodiversity and range-restricted endemism in karstic habitats rivals that of most other habitats throughout the tropics (see discussions in [1,4,5,10,56–61]). The sad irony is that, although these are some of the most imperiled ecosystems on the planet due to unregulated and unsustainable quarrying practices, only 1% of these terrains throughout Asia are afforded any form of legal protection. Therefore, the diversity of the karst-associated species in general—and *Cyrtodactylus* in particular—are, for the most part, without legal protection. Unfortunately, the immense financial returns from cement manufacturing makes the challenge of karst conservation difficult and many governments from developing nations that are willing to overlook sustainable quarrying policies in order to expand their economy [1]. Continued exploitation of karstic habitats for limestone shows no signs of abating.

5. Conclusions

This study echoes the results of Grismer et al. [6] in that karstic landscapes are exceedingly important for maintaining *Cyrtodactylus* diversity and serve as foci for their speciation and maintenance of their diversity. Referring to them as "imperiled arks of biodiversity" is somewhat misleading as these are ecological platforms for speciation that not only continue to generate the most speciose, independent, radiations of the Gekkota, but do so across a broad range of other taxonomic groups (e.g., [7–10,62]). Referring to them as "imperiled arks of biodiversity" instead of centers for speciation draws attention away from their importance as generators of biodiversity in an era of biodiversity crisis and could potentially lessen the urgency for legislative conservation measures.

Supplementary Materials: The following are available online at https://www.mdpi.com/article/10.3390/d13050183/s1, Table S1: Species, habitat preference with supporting references, species group designations, and GenBank accession numbers for specimens used in the SCM analysis. Species can be cross-referenced to Figure 6 by their GenBank no.

Author Contributions: Conceptualization, L.G., J.L.G. and P.L.W.J.; methodology, L.G., J.L.G. and P.L.W.J.; formal analysis, L.G. and P.L.W.J.; resources, N.A.P., M.D.L., S.K., S.C., C.S., S.Q., S.L. and J.C.; data curation, L.G. and P.L.W.J.; writing—original draft preparation, L.G.; writing—review and editing, all authors.; visualization, L.G.; supervision, L.G.; funding acquisition, S.K. and N.A.P. All authors have read and agreed to the published version of the manuscript.

Funding: This work was partially supported by the Nagao Natural Environment Foundation (2018-20) grant to Suranjan Karunarathna for field and lab works. This research was partially funded by the Russian Science Foundation grant number 19-14-00050 to Nikolay A. Poyarkov (sample collection, molecular and phylogenetic analyses). This research was funded by the Vietnam National Foundation for Science and Technology Development (NAFOSTED) under the grant number 106.05-2020.24 and the National Geographic grant number 230151 to Minh Duc Le.

Institutional Review Board Statement: Not applicable.

Informed Consent Statement: Not applicable.

Data Availability Statement: This paper follows all the relevant guidelines and ethics at https://www.mdpi.com/ethics.

Acknowledgments: We thank Steve Richards for additional information and photographs of taxa from New Guinea. N.A. Poyarkov thanks R.A. Nazarov, Valentina F. Orlova, Evgeniya N. Solovyeva and Eduard A. Galoyan (ZMMU, Russia), Andrei N. Kuznetsov and Svetlana P. Kuznetsova (JRVTC, Hanoi, Vietnam), Nikolai L. Orlov and Natalia B. Ananjeva (ZISP RAS, Russia), Parinya Pawangkhanant (AUP, Thailand), Than Zaw and May Thu Chit (Mandalay University, Myanmar), Peter Brakels (IUCN, Laos), Nguyen Van Tan (SVW, Vietnam), Indraneil Das (UNIMAS, Malaysia), Hung Ngoc Nguyen and Si-Min Lin (NTNU, Taiwan), Anna B. Vassilieva (IPEE RAS, Russia), Platon V. Yushchenko, Anna S. Dubrovskaya, Sabira S. Idiatullina, and Vladislav A. Gorin (MSU, Russia) for continuous support and assistance in the field and in the lab. S. Karunarathna thanks Kanishka Ukuwela, Anslem de Silva, Majintha Madawala, Dinesh Gabadage and Madhava Botejue for various support and assistance in the field and in the lab and the Department of Wildlife Conservation (WL/3/2/42/18a, b, c) and Forest Conservation Department (RandE/RES/NFSRCM/2019-04) for research permits. Chatmongkon Suwannapoom thanks the Unit of Excellence 2020 on Biodiversity and Natural Resources Management, University of Phayao (UoE63005). PLW's contribution to this paper constitutes contribution number 943 of the Auburn University Museum of Natural History.

Conflicts of Interest: The authors declare no conflict of interest.

References

1. Tolentino, P.J.; Navidad, J.R.L.; Angeles, M.D.; Fernandez, D.A.P.; Villanueva, E.L.C.; Obeña, R.D.R.; Buot, J.I.E. Review: Biodiversity of forests over limestone in Southeast Asia with emphasis on the Philippines. *Biodiversitas J. Biol. Divers.* **2020**, *21*, 1597–1613. [CrossRef]
2. Sodhi, N.S.; Koh, L.P.; Brook, B.W.; Ng, P.K.L. Southeast Asian biodiversity: An impending disaster. *Trends Ecol. Evol.* **2004**, *19*, 654–660. [CrossRef] [PubMed]
3. Sodhi, N.S.; Posa, M.R.C.; Lee, T.M.; Bickford, D.; Koh, L.P.; Brook, B.W. The state and conservation of Southeast Asian biodiversity. *Biodivers. Conserv.* **2010**, *19*, 317–328. [CrossRef]
4. Clements, R.; Ng, P.K.; Lu, X.X.; Ambu, S.; Schilthuizen, M.; Bradshaw, C.J. Using biogeographical patterns of endemic land snails to improve conservation planning for limestone karsts. *Biol. Conserv.* **2008**, *141*, 2751–2764. [CrossRef]
5. Clements, R.; Sodhi, N.S.; Schilthuizen, M.; Ng, P.K.L. Limestone karst of Southeast Asia: Imperiled arks of biodiversity. *Bioscience* **2006**, *56*, 733–742. [CrossRef]
6. Grismer, L.L.; Wood, P.L., Jr.; Le, M.D.; Quah, E.S.H.; Grismer, J.L. Evolution of habitat preference in 243 species of Bent–toed geckos (Genus *Cyrtodactylus* Gray, 1827) with a discussion of karst habitat conservation. *Ecol. Evol.* **2020**, *10*, 13717–13730. [CrossRef]
7. Barjadze, S.; Murvanidze, M.; Arabuli, T.; van Mumladze, L.; Pkhakadze, V.; Djanashvili, R.; Salakaia, M. *Annotated List of Inver-tebrates of the Georgian Karst Caves*; Georgian Academic Book: Tbilisi, Georgia, 2015; pp. 1–120.
8. Chen, W.H.; Zhang, Y.M.; Li, Z.Y.; Nguyen, Q.H.; Nguyen, T.H.; Shui, Y. *Hemiboea crystallina*, a new species of Gesneriaceae from karst regions of China and Vietnam. *Phytotaxa* **2018**, *336*, 23–56. [CrossRef]
9. Chin, S.C. The limestone hill flora of Malaya: Part 1. *Gard. Bull. Sing.* **1977**, *30*, 165–220.
10. Chung, W.-H.; Zhang, Y.-M.; Li, Z.-Y.; Nguyen, Q.-H.; Nguyen, T.-H.; Shui, Y.-M. Phylogenetic analyses of Begonia sect. *Coelocentrum* and allied limestone species of China shed light on the evolution of Sino-Vietnamese karst flora. *Bot. Stud.* **2014**, *55*, 1–35. [CrossRef]
11. Grismer, L.L.; Wood, P.L.; Poyarkov, N.A.; Le, M.D.; Kraus, F.; Agarwal, I.; Oliver, P.M.; Nguyen, S.N.; Nguyen, T.Q.; Karunarathna, S.; et al. Phylogenetic partitioning of the third-largest vertebrate genus in the world, *Cyrtodactylus* Gray, 1827 (Reptilia; Squamata; Gekkonidae) and its relevance to taxonomy and conservation. *Vert. Zool.* **2021**, *71*, 101–154. [CrossRef]
12. Uetz, P.; Freed, P.; Hošek, J. The Reptile Database. Available online: http://www.reptile-database.org (accessed on 14 February 2021).
13. Grismer, L.L.; Grismer, J.L. A re-evaluation of the phylogenetic relationships of the *Cyrtodactylus condorensis* group (Squamata; Gekkonidae) and a suggested protocol for the characterization of rock-dwelling ecomorphology in *Cyrtodactylus*. *Zootaxa* **2017**, *4300*, 486–504. [CrossRef]
14. Grismer, L.L.; Wood, P.L.; Thura, M.K.; Zin, T.; Quah, E.S.H.; Murdoch, M.L.; Grismer, M.S.; Lin, A.; Kyaw, H.; Lwin, N. Twelve new species of *Cyrtodactylus* Gray (Squamata: Gekkonidae) from isolated limestone habitats in east-central and southern Myanmar demonstrate high localized diversity and unprecedented microendemism. *Zool. J. Linn. Soc.* **2017**, *182*, 862–959. [CrossRef]
15. Hikida, T. Bornean Gekkonid Lizards of the Genus *Cyrtodactylus* (Lacertilia: Gekkonidae). *Jpn. J. Herpetol.* **1990**, *13*, 91–107. [CrossRef]
16. Johnson, C.B.; Evan Quah, S.H.; Anuar, S.; Muin, M.A.; Wood, J.P.L.; Grismer, J.L.; Greer, L.F.; Onn, C.K.; Ahmad, N.; Bauer, A.M.; et al. Phylogeography, geographic variation, and taxonomy of the Bent-toed Gecko *Cyrtodactylus quadrivirgatus* Taylor, 1962 from Peninsular Malaysia with the description of a new swamp dwelling species. *Zootaxa* **2012**, *3406*, 39–58. [CrossRef]
17. Kraus, F. Taxonomic partitioning of *Cyrtodactylus louisiadensis* (Lacertilia: Gekkonidae) from Papua New Guinea. *Zootaxa* **2008**, *1883*, 1–27. [CrossRef]
18. Welton, L.J.; Siler, C.D.; Diesmos, A.C.; Brown, R.M. Phylogeny-based species delimitation of southern Philippines bent-toed geckos and a new species of *Cyrtodactylus* (Squamata: Gekkonidae) from western Mindanao and the Sulu Archipelago. *Zootaxa* **2010**, *2390*, 49–68. [CrossRef]
19. Grismer, L.L. *Lizards of Peninsular Malaysia, Singapore and Their Adjacent Archipelagos*; Edition Chimaira: Frankfürt am Main, Germany, 2011; pp. 1–728.
20. Grismer, L.L.; Wood, P.L., Jr.; Thura, M.K.; Quah, E.S.; Murdoch, M.L.; Grismer, M.S.; Herr, M.W.; Lin, A.; Kyaw, H. Three more new species of *Cyrtodactylus* (Squamata: Gekkonidae) from the Salween Basin of eastern Myanmar underscore the urgent need for the conservation of karst habitats. *J. Nat. Hist.* **2018**, *52*, 1243–1294. [CrossRef]
21. Grismer, L.L.; Wood, P.L., Jr.; Quah, E.S.H.; Grismer, M.S.; Thura, M.K.; Oaks, J.R.; Lin, A. Two new species of *Cyrtodactylus* Gray, 1827 (Squamata: Gekkonidae) from a karstic archipelago in the Salween Basin of southern Myanmar (Burma). *Zootaxa* **2020**, *4718*, 151–183. [CrossRef]
22. Nielsen, S.V.; Oliver, P.M. Morphological and genetic evidence for a new karst specialist lizard from New Guinea (*Cyrtodactylus*: Gekkonidae). *R. Soc. Open Sci.* **2017**, *4*, 170781. [CrossRef]
23. Oliver, P.; Krey, K.; Mumpuni; Richards, S. A new species of bent-toed gecko (*Cyrtodactylus*, Gekkonidae) from the North Papuan Mountains. *Zootaxa* **2011**, *2930*, 22–32. [CrossRef]

24. Oliver, P.M.; Richards, S.J.; Mumpuni, M.; Rösler, H. The Knight and the King: Two new species of giant bent-toed gecko (*Cyrtodactylus*, Gekkonidae, Squamata) from northern New Guinea, with comments on endemism in the North Papuan Mountains. *ZooKeys* **2016**, *562*, 105–130. [CrossRef] [PubMed]
25. Ellis, M.; Pauwels, O.S.G. The bent-toed geckos (*Cyrtodactylus*) of the caves and karst of Thailand. *Cave Karst Sci.* **2012**, *39*, 16–22.
26. Ngo, V.T. Two new cave-dwelling species of *Cyrtodactylus* Gray (Squamata: Gekkonidae) from Southwestern Vietnam. *Zootaxa* **2008**, *1909*, 37–51. [CrossRef]
27. Nguyen, S.N.; Orlov, N.L.; Darevsky, I.S. Descriptions of two new species of the genus *Cyrtodactylus* Gray, 1827 (Squamata: Sauria: Gekkonidae) from southern Vietnam. *J. Herpetol.* **2006**, *13*, 215–226. [CrossRef]
28. Agarwal, I. Two new species of ground-dwelling *Cyrtodactylus* (Geckoella) from the Mysore Plateau, south India. *Zootaxa* **2016**, *4193*, 228–244. [CrossRef] [PubMed]
29. Grismer, L.L.; Wood, P.L., Jr.; Thura, M.K.; Win, N.M.; Quah, E.S.H. Two more new species of the *Cyrtodactylus peguensis* group (Squamata: Gekkonidae) from the fringes of the Ayeyarwady Basin, Myanmar. *Zootaxa* **2019**, *4577*, 274–294. [CrossRef] [PubMed]
30. Grismer, L.L.; Shahrul, A.; Quah, E.; Muin, M.A.; Chan, K.O.; Grismer, J.L.; Ahmad, N. A new spiny, prehensile-tailed species of *Cyrtodactylus* (Squamata: Gekkonidae) from Peninsular Malaysia with a preliminary hypothesis of relationships based on morphology. *Zootaxa* **2010**, *2625*, 40–52. [CrossRef]
31. Harvey, M.B.; O'connell, K.A.; Wostl, E.; Riyanto, A.; Kurniawan, N.; Smith, E.N.; Grismer, L.L. Redescription of *Cyrtodactylus lateralis* (Werner) (Squamata: Gekkonidae) and phylogeny of the prehensile-tailed *Cyrtodactylus*. *Zootaxa* **2016**, *4107*, 517–540. [CrossRef] [PubMed]
32. Dring, J.C.M. Amphibians and reptiles from northern Trengganu, Malaysia, with descriptions of two new geckos: *Cnemaspis* and *Cyrtodactylus*. *Bull. British Mus. (Nat. Hist.)* **1979**, *34*, 181–241.
33. Grismer, L.L. On the distribution and identification of *Cyrtodactylus brevipalmatus* Smith, 1923, and *Cyrtodactylus elok*, Dring, 1979. *Raffles Bull. Zool.* **2008**, *56*, 177–179. [CrossRef]
34. Grismer, L.L.; Wood, P.L., Jr.; Lim, K.K.P. *Cyrtodactylus majulah*, a new species of bent-toed gecko (Reptilia: Squamata: Gek-konidae) from Singapore and the Riau Archipelago. *Raffles Bull. Zool.* **2012**, *60*, 487–499.
35. Riyanto, A.; Grismer, L.L.; Wood, P.L., Jr. *Cyrtodactylus rosichonariefi* sp. nov. (Squamata: Gekkonidae), a new swamp-dwelling bent-toed gecko from Bunguran Island (Great Natuna), Indonesia. *Zootaxa* **2015**, *3964*, 114–124. [CrossRef] [PubMed]
36. Youmans, T.M.; Grismer, L.L. A new species of *Cyrtodactylus* (Reptilia: Squamata: Gekkonidae) from the Seribuat Archipelago, West Malaysia. *Herpetol. Nat. Hist.* **2006**, *10*, 61–70.
37. Geissler, P.; Hartmann, T.; Ihlow, F.; Neang, T.; Seng, R.; Wagner, P.; Böhme, W. Herpetofauna of the Phnom Kulen National Park, northern Cambodia—An annotated checklist. *Cambodian J. Nat. Hist.* **2019**, *1*, 40–63.
38. Wood, P.L., Jr.; Heinicke, M.P.; Jackman, T.R.; Bauer, A.M. Phylogeny of bent-toed geckos (*Cyrtodactylus*) reveals a west to east pattern of diversification. *Mol. Phylo. Evol.* **2012**, *65*, 992–1003. [CrossRef]
39. Hillis, D.M.; Moritz, C.; Mable, B.K.; Graur, D. *Molecular Systematics*, 2nd ed.; Sinauer Associates, Inc.: Sunderland, MA, USA, 1996; pp. 1–456. [CrossRef]
40. Katoh, M.; Kuma, M. MAFTT: A novel method for rapid sequence alignment based on fast Fourier transform. *Nucleic Acids Res.* **2002**, *30*, 3059–3066. [CrossRef] [PubMed]
41. Maddison, W.P.; Maddison, D.R. Mesquite: A Modular System for Evolutionary Analysis. Version 3.04. 2015. Available online: http://mesquiteproject.org (accessed on 23 December 2020).
42. Nguyen, L.-T.; Schmidt, H.A.; von Haeseler, A.; Minh, B.Q. IQ-TREE: A fast and effective stochastic algorithm for estimating maximum likelihood phylogenies. *Mol. Biol. Evol.* **2015**, *32*, 268–274. [CrossRef]
43. Trifinopoulos, J.; Nguyen, L.-T.; von Haeseler, A.; Minh, B.Q. W-IQ-TREE: A fast online phylogenetic tool for maximum likelihood analysis. *Nucleic Acids Res.* **2016**, *44*, W232–W235. [CrossRef]
44. Kalyaanamoorthy, S.; Minh, B.Q.; Wong, T.K.; von Haeseler, A.; Jermiin, L.S. Model Finder: Fast model selection for accurate phylogenetic estimates. *Nat. Meth.* **2017**, *14*, 587. [CrossRef]
45. Hoang, D.T.; Chernomor, O.; von Haeseler, A.; Minh, B.Q.; Vinh, L.S. UFBoot2: Improving the ultrafast bootstrap approximation. *Mol. Biol. Evol.* **2018**, *35*, 518–522. [CrossRef] [PubMed]
46. Minh, B.Q.; Nguyen, M.A.T.; von Haeseler, A. Ultrafast approximation for phylogenetic bootstrap. *Mol. Biol. Evol.* **2013**, *30*, 1188–1195. [CrossRef] [PubMed]
47. Bouckaert, R.; Heled, J.; Kühnert, D.; Vaughan, T.; Wu, C.-H.; Xie, D.; Suchard, M.A.; Rambaut, A.; Drummond, A.J. BEAST 2: A Software Platform for Bayesian Evolutionary Analysis. *PLoS Comput. Biol.* **2014**, *10*, e1003537. [CrossRef] [PubMed]
48. Miller, M.A.; Pfeiffer, W.; Schwartz, T. Creating the CIPRES Science Gateway for inference of large phylogenetic trees. In Proceedings of the Gateway Computing Environments Workshop (GCE), New Orleans, LA, USA, 14 November 2010; pp. 1–8.
49. Rambaut, A.; Suchard, M.A.; Xie, D.; Drummond, A.J. Tracer v1. 2014. Available online: https://bioweb.pasteur.fr/packages/pack@Tracer@v1.6 (accessed on 23 March 2021).
50. Rambaut, A.; Drummond, A.J. TreeAnnotator v1. 7.0. 2013. Available online: https://bioweb.pasteur.fr/packages/pack@Tracer@v1.6 (accessed on 23 March 2021).
51. Huelsenbeck, J.P.; Ronquist, F.; Nielsen, R.; Bollback, J.P. Bayesian inference of phylogeny and its impact on evolutionary biology. *Science* **2001**, *294*, 2310–2314. [CrossRef]

52. Wilcox, T.P.; Zwickl, D.J.; Heath, T.A.; Hillis, D.M. Phylogenetic relationships of the Dwarf Boas and a comparison of Bayesian and bootstrap measures of phylogenetic support. *Mol. Phylo. Evol.* **2002**, *25*, 361–371. [CrossRef]
53. Revell, L.J. Phytools: An R package for phylogenetic comparative biology (and other things). *Meth. Ecol. Evol.* **2012**, *3*, 217–223. [CrossRef]
54. Paradis, E.; Schilep, K. Ape 5.0: An environment for modern phylogenetics and evolutionary analyses in R. *Bioinformatics* **2018**, *35*, 526–528. [CrossRef]
55. Che, J.; Jiang, K.; Yan, F.; Zhang, Y.-P. *Amphibians and Reptiles of Tibet-Diversity and Evolution*; Science Press: Beijing, China, 2020; pp. 1–792. (In Chinese)
56. Do, T. Characteristics of karst ecosystems of Vietnam and their vulnerability to human impact. *Acta Geol. Sin.* **2002**, *75*, 325–329. [CrossRef]
57. Schilthuizen, M.; Liew, T.-S.; Bin Elahan, B.; Lackman-Ancrenaz, I. Effects of karst forest degradation on pulmonate and proso-branch land snail communities in Sabah, Malaysian Borneo. *Conserv. Biol.* **2005**, *19*, 949–954. [CrossRef]
58. Sterling, E.J.; Hurley, M.M.; Le, M.D. *Vietnam: A Natural History*; Yale University Press: New Haven, CT, USA, 2006; pp. 1–456.
59. Latinne, A.; Waengsothorn, S.; Herbreteau, V.; Michau, J.R.; Baillarguet, C. Thai limestone karsts: An impending biodiversity crisis. In Proceedings of the International Conference on Environment Supporting Food Energy Security: Crisis Opportunity, Bangkok, Thailand, 22–25 March 2001; pp. 176–187.
60. Grismer, L.L.; Wood, P.L., Jr.; Anuar, S.; Davis, H.R.; Cobos, A.J.; Murdoch, M.L. A new species of karst forest Bent-toed Gecko (genus *Cyrtodactylus* Gray) not yet threatened by foreign cement companies and a summary of Peninsular Malaysia's endemic karst forest herpetofauna and the need for its conservation. *Zootaxa* **2016**, *4061*, 1–17. [CrossRef] [PubMed]
61. Luo, Z.; Tang, S.; Jiang, Z.; Chen, J.; Fang, H.; Li, C. Conservation of terrestrial vertebrates in a global hotspot of karst in south-western China. *Sci. Rep.* **2016**, *6*, 25717. [CrossRef]
62. von Oheimb, P.V.; von Oheimb, K.C.M.; Hirano, T.; Do, V.T.; Luong, H.V.; Ablett, J.; Pham, S.V.; Naggs, F. Competition matters: Determining the drivers of land snail community assembly among limestone karst areas in northern Vietnam. *Ecol. Evol.* **2017**, *28*, 4136–4149. [CrossRef] [PubMed]

Article

Habitat Partitioning and Overlap by Large Lacertid Lizards in Southern Europe

Daniel Escoriza [1,*] and Félix Amat [2]

[1] GRECO, Institute of Aquatic Ecology, Campus Montillivi s/n, University of Girona, 17071 Girona, Spain
[2] Area d'Herpetologia, BiBIO, Museu de Granollers–Ciències Naturals, Palaudàries 102, 08402 Catalunya, Spain; felixamat09@gmail.com
* Correspondence: daniel_escoriza@hotmail.com

Abstract: South-western Europe has a rich diversity of lacertid lizards. In this study, we evaluated the occupancy patterns and niche segregation of five species of lacertids, focusing on large-bodied species (i.e., adults having >75 mm snout-vent length) that occur in south-western Europe (Italian to the Iberian Peninsula). We characterized the niches occupied by these species based on climate and vegetation cover properties. We expected some commonality among phylogenetically related species, but also patterns of habitat segregation mitigating competition between ecologically equivalent species. We used multivariate ordination and probabilistic methods to describe the occupancy patterns and evaluated niche evolution through phylogenetic analyses. Our results showed climate niche partitioning, but with a wide overlap in transitional zones, where segregation is maintained by species-specific responses to the vegetation cover. The analyses also showed that phylogenetically related species tend to share large parts of their habitat niches. The occurrence of independent evolutionary lineages contributed to the regional species richness favored by a long history of niche divergence.

Keywords: enhanced vegetation index; *Lacerta*; Mediterranean; niche partitioning; Sauria; *Timon*

Citation: Escoriza, D.; Amat, F. Habitat Partitioning and Overlap by Large Lacertid Lizards in Southern Europe. *Diversity* **2021**, *13*, 155. https://doi.org/10.3390/d13040155

Academic Editor: Francisco Javier Zamora-Camacho

Received: 17 March 2021
Accepted: 2 April 2021
Published: 4 April 2021

1. Introduction

Climate is a powerful environmental factor driving the process of niche diversification in reptiles [1,2]. Tolerance to maximum temperatures in reptiles is evolutionarily constrained, possibly because of the importance of external heat sources in maintaining activity and for bodily water balance [3,4]. For these reasons, there is a significant association between the composition of reptile assemblages and thermal latitudinal gradients [5]. However, the thermoregulation efficiency of reptiles is not only mediated by the overall climate conditions, but also by the temperature conditions in microhabitats [6]. Vegetation cover and structure regulate the patterns of reptile occurrence, but at a finer scale than the climate [7,8].

South-western Europe encompasses a relatively rich reptile fauna, favored by its topographic heterogeneity, insularity and mild climate conditions [9]. In this region, several groups of phylogenetically related species display complex patterns of overlap structured by environmental gradients or by interspecific interactions [10–12]. In this study we evaluated the niche occupancy patterns of five species of large lacertid lizard (i.e., adults having >75 mm snout-vent length) that occur in south-western Europe and comprise a monophyletic group [13]. We focused on phylogenetically related species because stronger competitive interactions among them can be expected [14]. The target species in this study included one species having a broad circumboreal distribution, *Lacerta agilis* (Linnaeus 1758), two western Mediterranean endemics *Lacerta bilineata* (Daudin, 1802) and *Timon lepidus* (Daudin, 1802), and two Iberian endemics *Lacerta schreiberi* (Bedriaga, 1878) and *Timon nevadensis* (Buchholz, 1963). Given these substantial chorological differences, it was also expected that these species would diverge in their environmental associations. For example, *T. nevadensis* mainly occupies semi-arid steppe-like habitats, *L. agilis* appears to be confined to sub-alpine meadows, and the other species are possibly more habitat generalists [15,16].

The purpose of this study was to investigate the patterns of habitat occupancy of these five species of large lacertids and evaluate them from an evolutionary perspective. We hypothesized that the diversity of large lacertid lizards in southern Europe would be favored by niche partitioning. However, this partitioning will be phylogenetically constrained because related species tend to occupy similar or equivalent niches [17]. To test these hypotheses, we used multivariate ordination and novel probabilistic methods that enabled quantification of the niche overlap between species, and the niche breadth.

2. Materials and Methods

2.1. Study Region and Surveys

The study region encompassed most of south-western Europe including the Iberian Peninsula, southern France and the Italian Peninsula (Figure 1). The region is dominated by Mediterranean climate types, ranging from subtropical warm desert to humid sub-Mediterranean and oceanic, and temperate to tundra-like types in the mountain ranges (Pyrenees, Alps) [18]. This environmental heterogeneity favors the concentration of high biotic diversity in this region, including species having xeric Mediterranean and meso-temperate affinities [19].

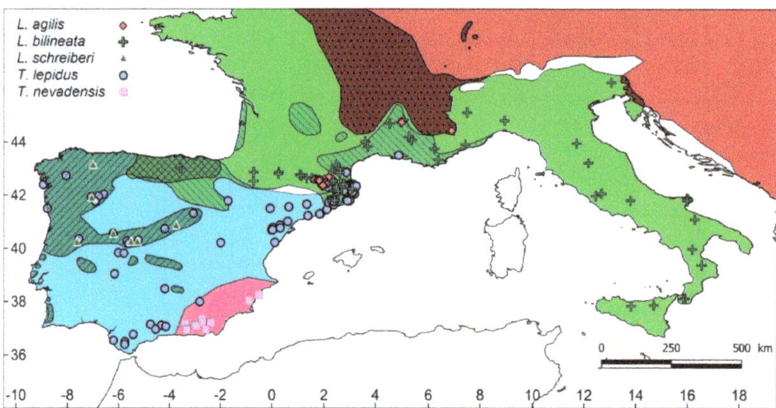

Figure 1. Map of the study region including the distribution of the species according to the IUCN (polygons) and the surveyed sites (circles). The polygons with diagonal stripes and dots indicate the areas of geographic overlap between two species or three species (grid).

Species records were obtained opportunistically, based on random habitat surveys conducted in the region from spring to autumn, following the annual activity periods. The surveys were planned to capture the maximum heterogeneity of habitats within the distribution ranges of these species, but with no a priori selection of the most suitable habitats (i.e., a random sampling of available habitats). In total, 823 sites were surveyed throughout south-western Europe by 1–3 observers; each site was visited only once. The occurrence of each species was assessed based on visual surveys and rock flipping, because both techniques have been used to build inventories of diurnal lizards [20]. The surveys were conducted on sunny days between 10:00 a.m. and 5:00 p.m. local time. The visual surveys were complemented with rock flipping in cases where identification of the species could not be visually ascertained. In total, 188 records of five species of large lacertids were obtained (Figure 1) and were distributed as follows: *L. agilis* (number of records = 29), *L. bilineata* (59), *L. schreiberi* (8), *T. lepidus* (80), and *T. nevadensis* (12).

2.2. Environmental Data

The niches for the species were described based on climate and vegetation. We used three variables to describe the climate: the maximum temperature of the warmest month, the minimum temperature of the coldest month, and the accumulated annual precipitation, provided by the WorldClim database [21]. These variables represent climatic properties that reportedly influence reptile ranges because they describe thermal extremes and available environmental moisture [22].

The influence of plant cover was assessed using a surrogate for vegetation primary productivity, the enhanced vegetation index (EVI; [23]); EVI values correlate positively with the density of trees [24]. The EVI data were obtained for the 2009–2019 period from the high-resolution (250 m pixel^{-1}) MODIS Collection EVI composite images [25]. The MODIS data were first checked to remove atmospheric artefacts, and then used to generate a series of variables describing the seasonal variability among habitats [26] including: the mean value (EVImean), the coefficient of variation (EVIcv), and the range (EVIrange), considering the inter-annual (mean value for 10 years) and spatial variability (mean value for 50 points, generated randomly within a maximum radius of 5 km). This larger area assessed the effect of the environment around the core habitat where specimens were found and took into account that species occurrence is sustained by isolated suitable habitat patch, but also by the interconnection of habitat patches that support the entire population [27].

2.3. Data Analysis

Species associations with two niche dimensions (climate and habitat) were visualized using outlying mean index (OMI) analysis [28]. The OMI ordination describes the species responses by quantifying their ecological marginality (the distance between the species centroid and the mean environmental conditions [28]). An OMI value close to zero indicates a higher similarity between the species position and the background environmental conditions. The OMI analysis also provided an estimate of the niche breadth of the species (tolerance index), and the proportion of the environmental variance explained by the OMI axes (residual tolerance; [28]). These analyses were carried out using the software package ade-4 [29] for R [30].

The niche overlap was estimated between pairs of species for a probabilistic niche region [31]. This overlap index was generated after 10,000 Monte Carlo draws for a niche region (alpha = 0.95), using the predictor variables. This analysis evaluates the probability that an individual of species X is found in the habitat of species Y, and vice versa, and produces two index values for a single pair of species (i.e., X$\rightarrow$Y and Y$\rightarrow$X) [32]. This method has the advantage of being weakly sensitive to the sample size [31], which is useful when evaluating the ecological overlap between species having dissimilar distributions, as was the case in this study. These analyses were carried out using the nicheROVER software package [32] for R.

We also compared the habitat characteristics between the pairs of species. Before modelling these associations, we tested the predictor variables for spatial autocorrelation using Moran's I correlograms [33]. Moran's I values were statistically significant, varying from 0.16 (EVIcv) to 0.55 (maximum temperature). To remove spatial autocorrelation, we built Binomial Generalized Linear Auto-Covariate Models (BGLAMs; [34]) selecting the best candidate model using the Akaike information criterion corrected for small sample sizes (AICc; [35]). In general, the best candidate models show lower AICc values, and an AICc weight $\geq$0.1 [35]. These analyses were carried out using the software packages spdep [36] and AICcmodavg [37] in the R environment.

2.4. Molecular Phylogenetics and Biogeography

Evolutionary relationships among the species of *Lacerta* and *Timon* were assessed by building a phylogenetic tree generated using Bayesian analysis of mitochondrial cytochrome b and 12s genes, obtained from GenBank (Supplementary Materials; Appendix I). The sequences were assembled and aligned using Bioedit 7.09 [38]. Our dataset comprised

32 sequences of variable length combining the two genes, and represented nine *Lacerta* and six *Timon* species, and four outgroups. The analysis of DNA evolution was conducted using jModelTest [39] and showed that the GTR+I+G model was the best for both the 12s and cytochrome b genes. We used three points of calibration to establish the times of divergence among species. The emergence of the Canary Island of El Hierro representing the divergence of *Gallotia caesaris caesaris* and *Gallotia caesaris gomerae* (1.0 Mya [40]) was used selecting a normal prior distribution (sigma = 0.02). The split between *Lacerta* and *Timon* was dated based using as a prior a gamma distribution based on a minimum age of 17.5 Mya [41] with shape and scale set to 1.0. The same prior distribution was used to calibrate the divergence between *L. viridis* and *L. bilineata* 8.7 Mya [42]. Bayesian analyses were performed using BEAST v 2.6.3. [43] running two chains of 5×10^8 iterations, sampling every 10,000 iterations. Chains were checked for convergence and ESS using Tracer 1.5 [44] and were combined after a burn-in of 99%. To reconstruct the biogeographic history of large lacertid lizards in south-western Europe we used ancestral range estimation using BioGeoBEARS implemented in RASP 4.2 [45]. The regions included were south-western Asia (Caucasus, Anatolia, Iran and the Middle East), Europe (excluding the Iberian Peninsula), north-western Africa and the Iberian Peninsula. The models of vicariance, dispersal and extinction allowing all the combinations of ancestral areas (except the Iberian Peninsula plus Asia) were evaluated using the Akaike criterion [46].

3. Results

The first two axes of the OMI explained 96.98% of total inertia (axis 1: 70.47%, axis 2: 26.51%). The first axis described a gradient from higher to lower temperature and precipitation and differentiated those species that occur under humid-cold conditions from those that occur under hot-dry conditions (Figure 2). The second axis described a transition between habitats having different vegetation cover, typically distinguishing habitats having a relatively high EVI and a low seasonal coefficient of variation (CV) (e.g., forests) from those having a relatively low EVI and a high seasonal CV (i.e., grasslands/cultivated lands; Figure 2). The genus *Timon* was separated from the genus *Lacerta* mainly along the first axis, the former showing a positive association with dry-warm climates (Figure 2). The niche indices indicated that a large part of the variation in the occurrence of species was explained by the environmental variables, with residual tolerance values between 19.2% (*L. agilis*) to 66.9% (*L. schreiberi*) (Table 1). In general, the species showed moderate to high distances from the environmental centroid, ranging from 20.2% (*L. schreiberi*) to 74.4% (*L. agilis*) (Table 1), which indicated that they occupied confined subspaces within the available environment (Figure 2). The tolerance indices consistently showed moderate to low values, ranging from 22.6% (*T. lepidus*) to 6.4% (*L. agilis*) (Table 1), indicating that these species differed in their niche sizes (Figure 2).

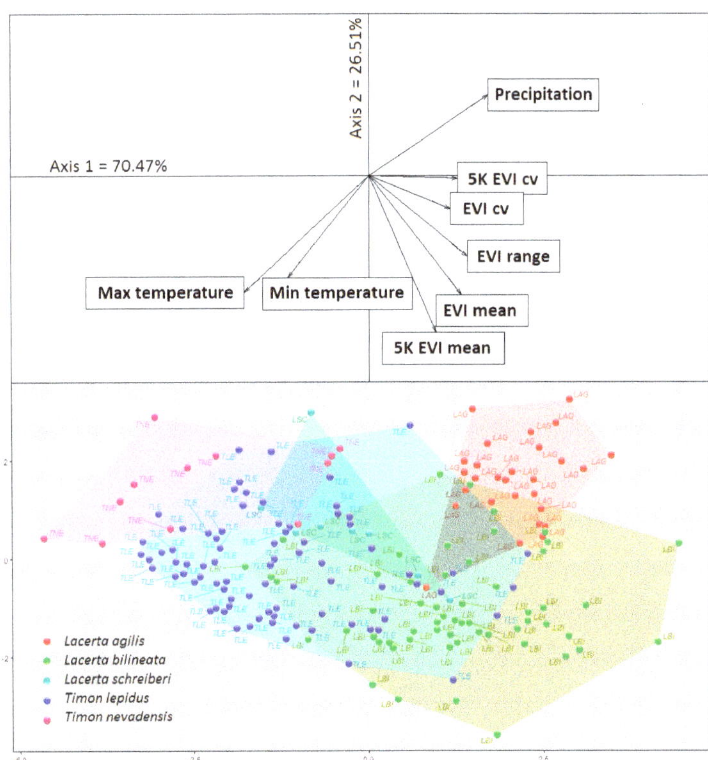

Figure 2. Outlying mean index (OMI) scatter plot. The lower panel shows the species sites, with a convex hull encompassing those that were conspecific. The upper panel shows the environmental factors represented as vectors. The plot origin (0,0) represents the average environmental conditions. LAG = *L. agilis*; LBI = *L. bilineata*; LSC = *L. schreiberi*; TLE = *T. lepidus*; TNE = *T. nevadensis*.

Table 1. Niche indices generated for the large lacertids in southwestern Europe. OMI, distance of the species' centroid to the average environmental conditions; Tolerance, niche breadth; Rtol, residual tolerance.

	OMI	Tol	Rtol
L. agilis	74.4	6.4	19.2
L. bilineata	30.3	18.3	51.5
L. schreiberi	20.2	12.8	66.9
T. lepidus	24.1	22.6	53.3
T. nevadensis	66.0	7.2	26.8

The niche overlap indices showed high values (>60) between pairs of sister species; for example, *T. nevadensis* → *T. lepidus*: 83.03 (Figures 3 and 4 and Table 2). Among phylogenetically more distant species the patterns were complex, including high (e.g., *L. schreiberi* → *L. bilineata*: 61.59; *L. bilineata* → *T. lepidus*: 86.03), moderate (*L. agilis* → *L. bilineata*: 53.54), and low (*L. bilineata* → *T. nevadensis*: 4.28) levels of overlap, and in some cases no overlap (*L. agilis* → *T. nevadensis*: 0.0) (Figure 3 and 4 and Table 2). Between some pairs of species, the overlap was highly asymmetric (e.g., *L. agilis-L. bilineata*, *T. lepidus-L. schreiberi*, and *T. lepidus-T. nevadensis*) indicating that the niche of species Y (smaller niche) was partially nested within that of species X (larger niche) (Figure 2).

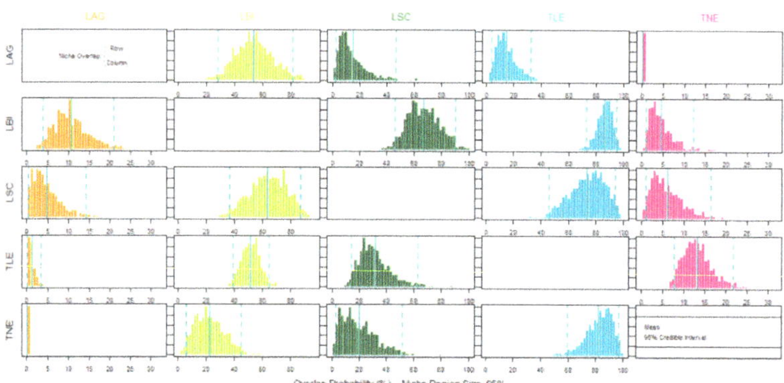

Figure 3. Ecological overlap between large lacertid species in southwestern Europe, estimated with the posterior distribution of the probabilistic niche overlap metric (%) for the niche region of alpha = 0.95. The posterior mean and 95% credible intervals are shown in sky blue lines.LAG = *L. agilis*; LBI = *L. bilineata*; LSC = *L. schreiberi*; TLE = *T. lepidus*; TNE = *T. nevadensis*.

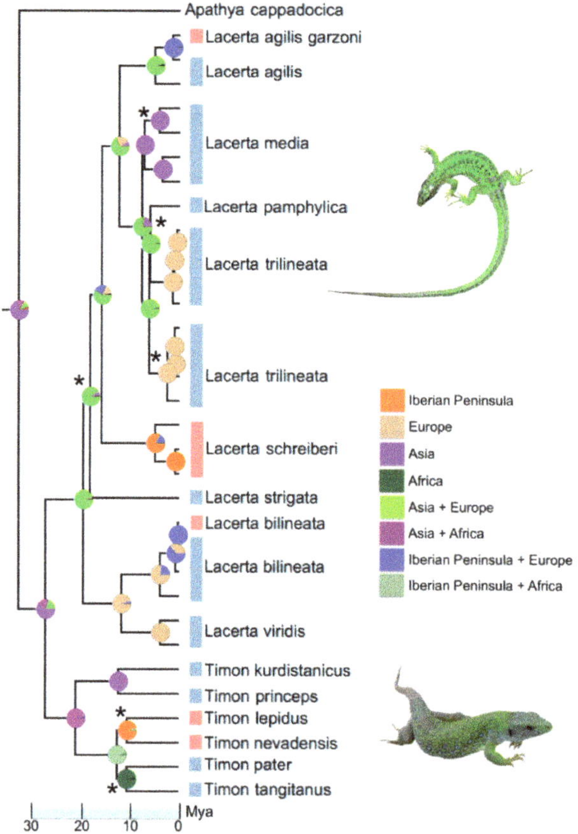

Figure 4. Phylogenetic relationships among *Lacerta* and *Timon* species estimated using Bayesian analysis on Cytochrome b and 12s mitochondrial genes. Asterisks denoted those nodes supported by posterior probabilities lower than 0.90. Pie charts depicted the probability of occurrence of the ancestors within the eight areas defined by the BioGeoBears analysis.

Table 2. Niche overlap between southwestern European large lacertid species. The overlap index described the probability that species X (row) appears in the habitat of the species Y (column), for a region $\alpha = 0.95$.

	L. agilis	*L. bilineata*	*L. schreiberi*	*T. lepidus*	*T. nevadensis*
L. agilis	–	53.54	14.48	14.76	0.0003
L. bilineata	10.43	–	65.56	86.03	4.28
L. schreiberi	4.54	61.59	–	71.93	5.69
T. lepidus	1.07	51.21	30.58	–	13.17
T. nevadensis	0.0	22.27	18.28	83.03	–

The BGLAMs showed than the niche separation between *L. agilis-L. bilineata* was mainly related to the by maximum-minimum temperatures ($R^2 = 0.257$) (Table 3). The separation between *L. agilis-T. lepidus* was also attributed to temperature, and to a lesser degree to plant cover ($R^2 = 0.469$). In contrast, the separation between *L. bilineata-T. lepidus* was mainly related to plant cover and to a lesser extent to the maximum temperature and annual precipitation ($R^2 = 0.181$). Comparisons that included *L. schreiberi* and *T. nevadensis* produced statistically poorly supported models and are not shown.

Table 3. Binomial Generalized Linear Auto-Covariate Models (BGLAMs) evaluating the environmental separation between pairs of species which geographically contact. *Lacerta schreiberi* and *T. nevadensis* were not included in the analyses. AIC, Akaike information criterion; AICWt, AIC weights. LAG = *L. agilis*; LBI = *L. bilineata*; TLE = *T. lepidus*. T, temperature; Prec, precipitations; EVI, Enhanced Vegetation Index; m, mean; cv, coefficient of variation.

	AIC	AICWt	R^2	Variables	Estimates
LAG-LBI	35.97	0.31	0.241	Tmin	−2.971
	37.33	0.16	0.257	Tmin Tmax	−2.288 −0.789
LAG-TLE	23.95	0.37	0.411	Tmax	−3.060
	25.80	0.15	0.416	Tmax Tmin	−2.726 −0.566
	26.31	0.11	0.469	Tmax Tmin EVIm	−2.408 −0.684 −1.534
LBI-TLE	63.31	0.30	0.144	EVIm	1.264
	64.64	0.16	0.154	EVIm EVIcv	1.286 0.275
	64.85	0.14	0.181	EVIm Tmax Prec	1.133 −0.658 0.301

The Bayesian phylogenetic analysis strongly supported the monophyly of the genera *Timon* and *Lacerta*, and most of the relationships across species within these genera (Figure 4). The most supported biogeographic model was DIVALIKE +J (AICc = 79.3, AICc weight = 0.630), indicating a founder event (j = 0.047) and low rates of dispersal and extinction (both <0.0001). Based on ancestral range reconstruction, large European lacertids arose in western Asia and independently colonized south-western Europe on four occasions. The oldest colonization event occurred approximately 8.9–15.5 Mya, during the invasion of the western Iberian Peninsula from Europe by the ancestor of *L. schreiberi*. The ancestor of *Timon* emigrated from western Asia to north-western Africa, and subsequently invaded the Iberian Peninsula approximately 6.5–13.0 Mya. The split of the ancestral Iberian species into *T. lepidus* and *T. tangitanus* probably occurred 4.6–11.5 Mya. The split of *L. bilineata* from the shared ancestor with *L. viridis* occurred approximately 8.7–9.4 Mya, possibly after their isolation in the Balkan and Italian peninsulas. *Lacerta bilineata* recently colonized the Iberian Peninsula from the Italian Peninsula (300,000–550,000 years ago) and the widespread palaearctic species *L. agilis*, colonized to the eastern Pyrenees from central Europe approximately 0.4–1.8 Mya.

4. Discussion

The OMI analysis showed that in south-western Europe the large lacertid lizards differed in niche marginality and tolerance. In general, environmental variables explained a major part of the variability in the occurrence of these species (residual tolerance 19.2% to 53.3%); an exception was *L. schreiberi* (residual tolerance 66.9%), possibly as a result of the low number of records. The ecological diversification of these species has possibly been driven by physiological tolerance, particularly variability in preferred temperature and the level of environmental moisture [47].

The analysis also indicated that *T. lepidus* and *L. bilineata* had the highest ecological tolerances, potentially enabling them to occupy parts of the niches of the other species. In contrast, *T. nevadensis* and *L. agilis*, occupied relatively narrow niches, at opposite extremes of the environmental range. Ecological partitioning was evident, but incomplete, between *L. agilis* and *L. bilineata* (maximum overlap probability 53.54%) and between *L. agilis-T. lepidus* (maximum overlap probability 14.76%). These species pairs consistently showed either geographical overlap (*L. agilis-L. bilineata* 170,990 km^2) or were almost completely parapatric (*L. agilis-T. lepidus*).

Regression models showed that the segregation between *L. agilis* and these species of lacertids (*L. bilineata* and *T. lepidus*) was mainly influenced by the variables describing temperature conditions, with the occurrence of *L. agilis* being negatively associated with temperature. This is consistent with the relict status of this species in the region, where it appears mainly isolated to mountain ranges, at altitudes above 1300 m [48]. The BGLAMs revealed that vegetation cover was unrelated to the niche partitioning between *L. agilis* and *L. bilineata*, possibly because of the use of open habitats in harsh subalpine environments by the later species [15,49].

There was wide but asymmetric overlap in the niches of *T. lepidus* and *L. bilineata*, and the niche of the latter was partially nested within that of *T. lepidus*. Both species also coexist over a wide geographic range (110,203 km^2), but *T. lepidus* is more widespread in the Iberian Peninsula. The later arrival of *L. bilineata* to the Iberian Peninsula may have been a disadvantage for this species, with it being excluded from potentially suitable habitats by the other lacertid species that evolved in this region (*T. lepidus, L. schreiberi*). Our results indicated that in the zone of coexistence between *T. lepidus* and *L. bilineata*, the species were largely segregated according to the vegetation cover. In general, forest habitats constitute unfavorable habitat for temperate lacertid lizards, which regulate their body temperature by sun basking [50]. The canopy of temperate forests greatly reduces light transmittance to the lower understory layers [51]. For this reason, closed forests are usually avoided by lacertid lizards, although they can colonize discontinuities in these habitats including path edges, interspersed meadows, forest margins and rocky outcrops [52–54]. *Lacerta bilineata* exploits microhabitats with a very dense vegetation cover, because this lizard thermoregulates efficiently using bushes and tree logs as basking platforms [47,55].

OMI and overlap analyses showed similarities in the patterns of habitat occupancy, between the species pairs *T. lepidus-T. nevadensis* and *L. schreiberi-L. bilineata*, and *T. nevadensis* niche was almost completely nested within that of *T. lepidus*. *Timon nevadensis* typically occupies relatively sparsely vegetated semiarid habitats, but *T. lepidus* is also able to exploit open, de-vegetated habitats in regions where *T. nevadensis* does not occur (e.g., in the shrub-steppes of the Ebro valley; [56]). These findings suggest that the range limits of both species may be sustained by interspecific interactions in the contact zones rather than by ecotonal transitions [57]. Our results showed a substantial habitat overlap between *L. schreiberi-L. bilineata*, which may trigger competitive interactions where both species contact geographically [58,59]. The results did not indicate that the species of Iberian origin (*L. schreiberi, T. lepidus*, and *T. nevadensis*) have greater habitat overlap than those of recent arrivals (e.g., *L. bilineata* and *L. agilis*). It is possible that during the prolonged isolation in the Iberian Peninsula these species diverged in their environmental niches, to reduce interspecific competition. Together these results indicate that the interaction of several mechanisms (i.e., interspecific competition, species evolutionary history, ecophysiological tolerance) determine the occurrence of large lizard species in the region, in a similar way to that observed for other lizard assemblages [60,61]. However, one limitation of our study was that it was not able to discern the relative importance of autoecological and synecological aspects of the distribution patterns.

5. Conclusions

The Iberian Peninsula has the richest lacertid fauna in south-western Europe, supporting five species of large lacertids. Our results revealed that this species richness is favored by climate niche partitioning, but with transitional areas of overlap where the segregation is maintained by species-specific responses to the level of vegetation cover. Interspecific competition may also play a key role in the patterns of occurrence, with the species that have arrived more recently on the Iberian Peninsula having been excluded from potentially favorable habitats because of prior habitat occupancy by species that evolved in this region. The occurrence of several independent evolutionary lineages has partly contributed to the species richness, which has been favored by a long history of ecological divergence between subclades having distinct geographic origins. The analyses applied in this study have the advantage of being weakly sensitive to sample size and are robust to spatially aggregated records, so they may be useful in disentangling the patterns of niche partitioning in assemblages including ecologically (or phylogenetically) related species structured by the effect of various interacting factors (i.e., competition, evolutionary history, environmental tolerance), which is characteristic of biotic communities in the Mediterranean region [62,63].

Supplementary Materials: The following are available online at https://www.mdpi.com/article/10.3390/d13040155/s1, Appendix I: Gene Bank accesion numbers and taxa selected for phylogenetic inference.

Author Contributions: Conceptualization, D.E. and F.A.; methodology, D.E. and F.A.; software, D.E. and F.A.; validation, D.E. and F.A.; formal analysis, D.E. and F.A.; investigation, D.E. and F.A.; resources, D.E.; writing—original draft preparation, D.E. and F.A.; writing—review and editing, D.E. and F.A. All authors have read and agreed to the published version of the manuscript.

Funding: This research received no external funding.

Institutional Review Board Statement: Not applicable.

Informed Consent Statement: Not applicable.

Data Availability Statement: The authors confirm that the data supporting the findings of this study are available within the article [and/or] its Supplementary Materials.

Acknowledgments: The authors thank Jihène Ben Hassine, Guillem Pascual, Alberto Sánchez-Vialas, and Roberto López for their assistance during the field work.

Conflicts of Interest: The authors declare no conflict of interest.

References

1. Jezkova, T.A.; Wiens, J.J. Testing the role of climate in speciation: New methods and applications to squamate reptiles (lizard and snakes). *Mol. Ecol.* **2018**, *27*, 2754–2769. [CrossRef]
2. Velasco, J.A.; Martínez-Meyer, E.; Flores-Villela, O.; García, A.; Algar, A.C.; Köhler, G.; Daza, J.M. Climatic niche attributes and diversification in *Anolis* lizards. *J. Biogeogr.* **2016**, *43*, 134–144. [CrossRef]
3. Rozen-Rechels, D.; Farigoule, P.; Agostini, S.; Badiane, A.; Meylan, S.; Le Galliard, J.F. Short-term change in water availability influences thermoregulation behaviours in a dry-skinned ectotherm. *J. Anim. Ecol.* **2020**, *89*, 2099–2110. [CrossRef]
4. Grigg, J.W.; Buckley, L.B. Conservatism of lizard thermal tolerances and body temperatures across evolutionary history and geography. *Biol. Lett.* **2013**, *29*, 20121056. [CrossRef]
5. Rodríguez, M.Á.; Belmontes, J.A.; Hawkins, B.A. Energy, water and large-scale patterns of reptile and amphibian species richness in Europe. *Acta Oecologica* **2005**, *28*, 65–70. [CrossRef]

6. Law, B.S.; Bradley, R.A. Habitat use and basking site selection in the water skink, *Eulamprus quoyii*. *J. Herpetol.* **1990**, *24*, 235–240. [CrossRef]
7. Capula, M.; Luiselli, L.; Rugiero, L. Comparative ecology in sympatric *Podarcis muralis* and *P. sicula* (Reptilia: Lacertidae) from the historical centre of Rome: What about competition and niche segregation in an urban habitat? *Ital. J. Zool.* **1993**, *60*, 287–291.
8. Escoriza, D. Patterns of alpha diversity among Tunisian lizards (Lacertidae). *J. Arid Environ.* **2018**, *151*, 23–30. [CrossRef]
9. Speybroeck, J.; Beukema, W.; Bok, B.; Van Der Voort, J. *Field Guide to the Amphibians and Reptiles of Britain and Europe*; Bloomsbury: London, UK, 2016.
10. Arnold, E.N. Resource partition among lacertid lizards in southern Europe. *J. Zool. Lond. B* **1987**, *1*, 739–782. [CrossRef]
11. Pollo, C.J.; Pérez-Mellado, V. An analysis of a Mediterranean assemblage of three small lacertid lizards in central Spain. *Acta Oecologica* **1991**, *12*, 655–671.
12. Escoriza, D.; Pascual, G.; Sánchez-Vialas, A. Habitat use in south-west European skinks (genus *Chalcides*). *PeerJ* **2018**, *6*, e4274. [CrossRef] [PubMed]
13. Tonini, J.F.R.; Beard, K.H.; Ferreira, R.B.; Jetz, W.; Pyron, R.A. Fully-sampled phylogenies of squamates reveal evolutionary patterns in threat status. *Biol. Conserv.* **2016**, *204*, 23–31. [CrossRef]
14. Labra, A.; Pienaar, J.; Hansen, T.F. Evolution of thermal physiology in *Liolaemus* lizards: Adaptation, phylogenetic inertia, and niche tracking. *Am. Nat.* **2009**, *174*, 204–220. [CrossRef] [PubMed]
15. Amat, F.A.; Llorente, G.A.; Carretero, M.A. A preliminary study on thermal ecology, activity times and microhabitat use of *Lacerta agilis* (Squamata: Lacertidae) in the Pyrenees. *J. Zool.* **2003**, *52*, 413–422.
16. Salvador, A. *Reptiles, 2ª Edición Revisada y Aumentada*; Fauna Ibérica; MNCN-CSIC: Madrid, Spain, 2014; Volume 10.
17. Wiens, J.J.; Graham, C.H. Niche conservatism: Integrating evolution, ecology, and conservation biology. *Annu. Rev. Ecol. Evol. Syst.* **2005**, *36*, 519–539. [CrossRef]
18. Kriticos, D.J.; Webber, B.L.; Leriche, A.; Ota, N.; Macadam, I.; Bathols, J.; Scott, J.K. CliMond: Global high-resolution historical and future scenario climate surfaces for bioclimatic modelling. *Methods Ecol. Evol.* **2012**, *3*, 53–64. [CrossRef]
19. Cuttelod, A.; García, N.; Abdul Malak, D.; Temple, H.; Katariya, V. The Mediterranean: A Biodiversity Hotspot under Threat. In *The 2008 Review of The IUCN Red List of Threatened Species*; Vié, J.-C., Hilton-Taylor, C., Stuart, S.N., Eds.; IUCN: Gland, Switzerland, 2020.
20. Losos, J.B.; Marks, J.C.; Schoener, T.W. Habitat use and ecological interactions of an introduced and a native species of *Anolis* lizard on Grand Cayman, with a review of the outcomes of anole introductions. *Oecologia* **1993**, *95*, 525–532. [CrossRef]
21. Fick, S.E.; Hijmans, R.J. WorldClim 2: New 1-km spatial resolution climate surfaces for global land areas. *Int. J. Climatol.* **2017**, *37*, 4302–4315. [CrossRef]
22. Araújo, M.B.; Thuiller, W.; Pearson, R.G. Climate warming and the decline of amphibians and reptiles in Europe. *J. Biogeogr.* **2006**, *33*, 1712–1728. [CrossRef]
23. Huete, A.; Didan, K.; Miura, T.; Rodriguez, E.P.; Gao, X.; Ferreira, L.G. Overview of the radiometric and biophysical performance of the MODIS vegetation indices. *Remote Sens. Environ.* **2002**, *83*, 195–213. [CrossRef]
24. Waring, R.H.; Coops, N.C.; Fan, W.; Nightingale, J.M. MODIS enhanced vegetation index predicts tree species richness across forested ecoregions in the contiguous USA. *Remote Sens. Environ.* **2006**, *103*, 218–226. [CrossRef]
25. Didan, K. MOD13Q1 Modis/Terra Vegetation Indices 16-Day L3 Global 250m SIN Grid V006. Nasa Eosdis Land Processes DAAC. Available online: https://modis.ornl.gov (accessed on 10 October 2020).
26. de Oliveira, B.S.; Ferreira, M.E.; Coutinho, A.C.; Esquerdo, J.C. Phenological-metric algorithm for mapping soybean in savanna biome in Brazil. *Aust. J. Crop Sci.* **2019**, *13*, 1456–1466. [CrossRef]
27. Templeton, A.R.; Brazeal, H.; Neuwald, J.L. The transition from isolated patches to a metapopulation in the eastern collared lizard in response to prescribed fires. *Ecology* **2011**, *92*, 1736–1747. [CrossRef] [PubMed]
28. Dolédec, S.; Chessel, D.; Gimaret-Carpentier, C. Niche separation in community analysis: A new method. *Ecology* **2000**, *81*, 2914–2927. [CrossRef]
29. Dray, S.; Dufour, A.B. The ade4 package: Implementing the duality diagram for ecologists. *J. Stat. Softw.* **2007**, *22*, 1–20. [CrossRef]
30. R Development Core Team. *R: A Language and Environment for Statistical Computing*; R Foundation for Statistical Computing: Vienna, Austria, 2020. Available online: https://www.R-project.org (accessed on 10 September 2020).
31. Swanson, H.K.; Lysy, M.; Power, M.; Stasko, A.D.; Johnson, J.D.; Reist, J.D. A new probabilistic method for quantifying n-dimensional ecological niches and niche overlap. *Ecology* **2015**, *96*, 318–324. [CrossRef] [PubMed]
32. Lysy, M.; Stasko, A.D.; Swanson, H.K. nicheROVER: (Niche) (R)Egion and Niche (Over)Lap Metrics for Multidimensional Ecological Niches. R Package Version 1.0. Available online: https://CRAN.R-project.org/package=nicheROVER (accessed on 4 December 2020).
33. Legendre, P. Spatial autocorrelation: Trouble or new paradigm? *Ecology* **1993**, *7*, 1659–1673. [CrossRef]
34. Waller, L.A. Spatial Models for Categorical Data. In *Wiley StatsRef: Statistics Reference Online*; Wiley: New York, NY, USA, 2014.
35. Burnham, K.P.; Anderson, D.R. A practical information-theoretic approach. In *Model Selection and Multimodel Inference*, 2nd ed.; Springer: New York, NY, USA, 2002.
36. Bivand, R.S.; Wong, D.W. Comparing implementations of global and local indicators of spatial association. *Test* **2018**, *27*, 716–748. [CrossRef]

37. Mazerolle, M.J. AICcmodavg: Model Selection and Multimodel Inference Based on (Q)AIC(c). R Package Version 2.3-1. Available online: https://cran.r-project.org/package=AICcmodavg (accessed on 6 January 2021).
38. Hall, T.A. BioEdit: A user-friendly biological sequence alignment editor and analysis program for Windows 95/98/NT. *Nucleic Acids Symp. Ser.* **1999**, *41*, 95–98.
39. Posada, D. jModelTest: Phylogenetic model averaging. *Mol. Biol. Evol.* **2008**, *25*, 1253–1256. [CrossRef]
40. Carranza, S.; Arnold, E.N.; Amat, F. DNA phylogeny of *Lacerta* (*Iberolacerta*) and other lacertine lizards (Reptilia: Lacertidae): Did competition cause long-term mountain restriction? *Syst. Biod.* **2004**, *2*, 57–77. [CrossRef]
41. Cernansky, A. Earliest world record of green lizards (Lacertilia, Lacertidae) from the Lower Miocene of Central Europe. *Biologia* **2010**, *65*, 737–741. [CrossRef]
42. Venczel, M. Lizards from the late Miocene of Polgárdi (W-Hungary). *Nymphaea* **2006**, *33*, 25–38.
43. Bouckaert, R.; Vaughan, T.G.; Barido-Sottani, J.; Duchêne, S.; Fourment, M.; Gavryushkina, A.; Heled, J.; Jones, G.; Kühnert, D.; De Maio, N.; et al. BEAST 2.5: An advanced software platform for Bayesian evolutionary analysis. *PLoS Comp. Biol.* **2019**, *15*, e1006650. [CrossRef]
44. Rambaut, A.; Drummond, A.J. Tracer, Version 1.4. 2007. Available online: http://beast.bio.ed.ac.uk/Tracer (accessed on 10 September 2015).
45. Yu, Y.; Harris, A.J.; Blair, C.; He, X. RASP (Reconstruct Ancestral State in Phylogenies): A tool for historical biogeography. *Mol. Phyl. Evol.* **2015**, *87*, 46–49. [CrossRef]
46. Akaike, H. A new look at the statistical model identification. *IEEE Tran. Autom. Cont.* **1974**, *19*, 716–723. [CrossRef]
47. Saint Girons, M.C. Relations interspécifiques et cycle d'activité chez *Lacerta viridis* et *Lacerta agilis* (Sauria, Lacertidae). *Vie et Milieu* **1976**, *26*, 115–132.
48. Geniez, P.; Cheylan, M. *Les Amphibiens et Les Reptiles du Languedoc-Roussillon et Régions Limitrophes: Atlas Biogéographique*; Biotope: Mèze, France, 2012.
49. Arribas, O.J. Distribución y estatus de *Lacerta agilis* y *Zootoca vivipara* en Cataluña. *Bull. Soc. Catalana Herpetol.* **1999**, *14*, 10–21.
50. House, S.M.; Taylor, P.J.; Spellerberg, I.F. Patterns of daily behaviour in two lizard species *Lacerta agilis* L. and *Lacerta vivipara* Jacquin. *Oecologia* **1979**, *44*, 396–402. [CrossRef] [PubMed]
51. Sercu, B.K.; Baeten, L.; van Coillie, F.; Martel, A.; Lens, L.; Verheyen, K.; Bonte, D. How tree species identity and diversity affect light transmittance to the understory in mature temperate forests. *Ecol. Evol.* **2017**, *7*, 10861–10870. [CrossRef] [PubMed]
52. García, J.D.D.; Arévalo, J.R.; Fernández-Palacios, J.M. Road edge effect on the abundance of the lizard *Gallotia galloti* (Sauria: Lacertidae) in two Canary Islands forests. *Biodivers. Conserv.* **2007**, *16*, 2949–2963. [CrossRef]
53. Amo, L.; López, P.; Martín, J. Natural oak forest vs. ancient pine plantations: Lizard microhabitat use may explain the effects of ancient reforestations on distribution and conservation of Iberian lizards. *Biodivers. Conserv.* **2007**, *16*, 3409–3422. [CrossRef]
54. Schiavo, R.M.; Ferri, V. Ciclo annuale di *Lacerta bilineata* (Daudin, 1802) nella pianura padana lombarda. *Pianura* **1996**, *10*, 5–12.
55. Nettmann, H.K.; Rykena, S. *Lacerta viridis* (Laurenti, 1768)—Smaragdeidechse. In *Handbuch der Reptilien und Amphibien Europas*; Band 2/I. Echsen II (Lacerta); Bohme, W., Ed.; Aula-Verlag: Wiesbaden, Germany, 1984; pp. 129–189.
56. Suárez, F.; Manrique, J. Low breeding success in Mediterranean shrubsteppe passerines: Thekla lark *Galerida theklae*, lesser short-toed lark *Calandrella rufescens*, and black-eared wheatear *Oenanthe hispanica*. *Ornis Scand.* **1992**, *23*, 24–28. [CrossRef]
57. Miraldo, A.; Faria, C.; Hewitt, G.M.; Paulo, O.S.; Emerson, B.C. Genetic analysis of a contact zone between two lineages of the ocellated lizard (*Lacerta lepida* Daudin 1802) in south-eastern I beria reveal a steep and narrow hybrid zone. *J. Zool. Syst. Evol. Res.* **2013**, *51*, 45–54. [CrossRef]
58. Barbadillo, L.J. Nuevas citas herpetológicas para la Provincia de Burgos. *Rev. Esp. Herpetol.* **1986**, *1*, 57–61.
59. Delibes, A.; Salvador, A. Censos de lacértidos en la cordillera Cantábrica. *Rev. Esp. Herpetol.* **1986**, *1*, 335–361.
60. Barbault, R.; Stearns, S. Towards an evolutionary ecology linking species interactions, life-history strategies and community dynamics: An introduction. *Acta Oecologica* **1991**, *12*, 3–10.
61. Spawls, S.; Rotich, D. An annotated checklist of the lizards of Kenya. *J. East Afr. Nat. Hist.* **1997**, *86*, 61–83. [CrossRef]
62. Thompson, J.D.; Lavergne, S.; Affre, L.; Gaudeul, M.; Debussche, M. Ecological differentiation of Mediterranean endemic plants. *Taxon* **2005**, *54*, 967–976. [CrossRef]
63. Hampe, A.; Arroyo, J.; Jordano, P.; Petit, R.J. Rangewide phylogeography of a bird-dispersed Eurasian shrub: Contrasting Mediterranean and temperate glacial refugia. *Mol. Ecol.* **2003**, *12*, 3415–3426. [CrossRef] [PubMed]

 MDPI

Article

Does Hyperoxia Restrict Pyrenean Rock Lizards *Iberolacerta bonnali* to High Elevations?

Eric J. Gangloff [1,*], Sierra Spears [1], Laura Kouyoumdjian [2], Ciara Pettit [1] and Fabien Aubret [2]

[1] Department of Zoology, Ohio Wesleyan University, Delaware, OH 43015, USA; ssspears@owu.edu (S.S.); clpettit@owu.edu (C.P.)
[2] Station d'Ecologie Théorique et Expérimentale du CNRS–UPR 2001, 09200 Moulis, France; laura.kouyoumdjian@hotmail.fr (L.K.); fabien.aubret@sete.cnrs.fr (F.A.)
* Correspondence: ejgangloff@owu.edu

Abstract: Ectothermic animals living at high elevation often face interacting challenges, including temperature extremes, intense radiation, and hypoxia. While high-elevation specialists have developed strategies to withstand these constraints, the factors preventing downslope migration are not always well understood. As mean temperatures continue to rise and climate patterns become more extreme, such translocation may be a viable conservation strategy for some populations or species, yet the effects of novel conditions, such as relative hyperoxia, have not been well characterised. Our study examines the effect of downslope translocation on ectothermic thermal physiology and performance in Pyrenean rock lizards (*Iberolacerta bonnali*) from high elevation (2254 m above sea level). Specifically, we tested whether models of organismal performance developed from low-elevation species facing oxygen restriction (e.g., hierarchical mechanisms of thermal limitation hypothesis) can be applied to the opposite scenario, when high-elevation organisms face hyperoxia. Lizards were split into two treatment groups: one group was maintained at a high elevation (2877 m ASL) and the other group was transplanted to low elevation (432 m ASL). In support of hyperoxia representing a constraint, we found that lizards transplanted to the novel oxygen environment of low elevation exhibited decreased thermal preferences and that the thermal performance curve for sprint speed shifted, resulting in lower performance at high body temperatures. While the effects of hypoxia on thermal physiology are well-explored, few studies have examined the effects of hyperoxia in an ecological context. Our study suggests that high-elevation specialists may be hindered in such novel oxygen environments and thus constrained in their capacity for downslope migration.

Keywords: high elevation; hyperoxia; sprint performance; thermal performance curve; thermal preference

Citation: Gangloff, E.J.; Spears, S.; Kouyoumdjian, L.; Pettit, C.; Aubret, F. Does Hyperoxia Restrict Pyrenean Rock Lizards *Iberolacerta bonnali* to High Elevations?. *Diversity* **2021**, *13*, 200. https://doi.org/10.3390/d13050200

Academic Editors: Francisco Javier Zamora-Camacho, Mar Comas and Luc Legal

Received: 20 March 2021
Accepted: 5 May 2021
Published: 11 May 2021

Publisher's Note: MDPI stays neutral with regard to jurisdictional claims in published maps and institutional affiliations.

1. Introduction

Mountains cover approximately 30% of the world's land surface [1]. These biodiversity hotspots [2] harbour virtually all life forms (including diversity of bacteria [3,4], insects [5,6], arachnids [7], gastropods [8,9], fish [10,11], amphibians [12,13], mammals [14,15], birds [16], and squamate reptiles [17,18]). Mountain ecological landscapes are characterised by altitudinal zonation [19], where organisms tend to be adapted to a relatively narrow range of environmental conditions including colder temperature regimes (mean and extremes), strong UV irradiance, and lower atmospheric pressure, thus reduced oxygen availability as altitude increases. Although examples abound where geographically widespread species usually constrained to low elevation areas have successfully established along parts of the elevational gradient [20–23], plants and animals found in high altitudinal zones tend to become isolated since the conditions above and below a particular zone will be inhospitable and thus restrict their movements or dispersal. In extreme examples, such isolated ecological systems have been coined sky or continental islands [16,24,25].

Physiological adaptations to high-elevation life have attracted considerable scientific attention, including in humans, domesticated animals, and wild animal populations [26–28], notably in relation to colder temperature regimes, UV irradiance, and lower oxygen availability. However, it is less clear how such adaptations may prevent or hinder population movement, especially toward lower elevations. In other words, are populations found at high elevation effectively restricted to this elevation, and if so, why? Several non-exclusive hypotheses exist. Most obviously, environmental conditions below the current elevation may exceed the organisms' tolerance (i.e., beyond its fundamental niche), being too hot or too dry for instance [29–31]. Alternatively, lower-elevation niches may be exploited by a direct competitor, harbour a predator, or may lack a suitable food source [32–34].

One such example comes from an endemic trio of lizard species in the genus *Iberolacerta*, namely, *I. aranica*, *I. aurelioi*, and *I. bonnali*. These three species exhibit non-overlapping distributions between 1500 and 3100 m above sea level (ASL) in the Pyrénées mountains of southwestern Europe (France, Andorra, and Spain). They occur as a constellation of small populations, with high degrees of genetic isolation amongst populations of the three species [35,36] presumably due to very low dispersal rates amongst mountain peaks [37]. In the case of *Iberolacerta*, it was suggested that their restricted distribution resulted either (1) from their cold-adapted thermal physiology (i.e., low tolerance for high temperature, resulting in a reduction in their activity budgets by excess of heat [38–42]) or (2) from competitive exclusion from wall lizards (*Podarcis* spp. [43]). In support of the latter, some studies suggest that competition with *Podarcis* might affect the presence of high-elevation specialist *Iberolacerta* spp. through antagonistic interactions and competition for access to preferred thermal habitat ([44,45], but see [46]).

Recent studies [47,48] suggest that *Podarcis* may be suited to higher-elevation colonisation beyond its current range due to embryonic developmental resilience to lowered oxygen availability when transplanted to high elevation (≈3000 m ASL), well above its maximum recorded elevation (i.e., 2200 m ASL [35,49]). Moreover, *Podarcis* is locally observed to expand its range upslope at a steady but rapid pace in the Pyrénées [49], suggesting that fast colonisation might occur in the coming decades. This will inevitably bring more *Podarcis* into contact with *Iberolacerta* and foster potential competition for territories, nesting sites, and food, as well as potentially exposing *Iberolacerta* to novel diseases and parasites. Current climate change will only facilitate this process [29,50,51]: high-altitude areas are warming faster than the global average [52–54], and *Podarcis* are a thermophilic species successful at establishing in new environments [43,55–58].

With this study, we tested an additional, non-exclusive hypothesis of the mechanism limiting *Iberolacerta* to high elevation: we propose that *Iberolacerta* species have adapted to high elevation hypoxia to a point where sea-level oxygen levels (hyperoxia, from *Iberolacerta*'s perspective) may hinder organismal function. As an analogy, the metabolic cold adaptation hypothesis (MCA) predicts that ectotherms from colder environments (higher latitudes or elevations) will have elevated metabolic rates compared to those from warmer climates at a given temperature [59]. Increased metabolic rates are predicted to be adaptive by allowing accelerated physiological processes in environments that feature shorter periods of optimal conditions [60,61]. On the other hand, such adaptations become rapidly detrimental (i.e., metabolically very costly) if environment temperature increases (e.g., via climate change or dispersal). In the same manner, organisms adapted to maintaining organismal function in low-oxygen conditions may suffer under conditions of increased oxygen availability. For example, this may disrupt oxidative phosphorylation pathways that can either reduce the efficiency of aerobic metabolism or result in the production of potentially harmful byproducts [62–64].

To test this hypothesis, we studied the effect of translocation to low elevation on ectothermic thermal physiology and performance. Shifts in elevation most notably affect the total partial pressure exhibited by the atmosphere, which will reduce oxygen availability at high elevations and increase availability at low elevations. Utilising Pyrenean rock lizards

(*Iberolacerta bonnali*), we measured traits known to have important consequences for both fitness and thermally-dependent physiological processes: sprint speed and preferred body temperature. If high-elevation specialists are able to process increased oxygen when available, we predicted both sprint speed and preferred body temperatures will increase, in the opposite direction from what has been observed in low-elevation organisms brought into hypoxia [65–67]. In contrast, under our novel proposal that hyperoxia limits high-elevation species from moving downslope, we predicted performance decrements and reduced preferred body temperatures after transplanting lizards to low elevation. Thus, we sought to understand how organisms are adapted to their specific oxygen environment and their potential behavioural and physiological responses to novel environments. Quantifying these responses is essential in addressing the question of whether abiotic factors, such as oxygen availability, represent absolute constraints on organismal performance or whether organisms are specifically adapted to the resources in their environment and any deviance from these levels—either increase or decrease—can restrict physiological processes and performance.

2. Materials and Methods

2.1. Study Species

The Pyrenean rock lizard (*Iberolacerta bonnali* Lantz, 1927; Figure 1) is a diurnal, heliothermic species endemic to the alpine and subalpine environments of the Pyrénées Mountains [35] and can be found at elevations between 1550 and 3062 m ASL [68]. Its annual period of activity is very short due to cold temperatures and the presence of snow most of the year [35,68,69]. This restricts their reproductive cycle to one clutch per year with an average of three eggs [70]. Being a highly endemic patrimonial species with a very restricted range, it is listed on the IUCN red list of threatened reptile species in Europe [71].

Figure 1. Two adult *Iberolacerta bonnali* basking in their natural environment. Photograph by Fabien Aubret.

2.2. Experimental Design

Since female reproductive status cannot always be ascertained (i.e., early vitellogenesis) and because carrying eggs may affect performance and thermoregulatory patterns in squamates [72–74], the study was carried out with male lizards only. We captured 40 male lizards around the Lac d'Oncet at 2254 m ASL (Department of Hautes-Pyrénées, France) between 2 and 7 July 2020 using the lasso method [75,76] during peak activity hours (09 h 30–15 h 30). Immediately after capture (within 10 s), body temperature was recorded with an infrared thermometer from a 30 cm distance (infrared thermometer Trotec BP21,

Marchtrenk, Austria, distance:measurement spot ratio 12:1). On the day of capture, we measured the body mass with a precision scale to the nearest 0.01 g (range: 1.24–4.28; mean ± s.d.: 2.64 ± 0.83 g) and the snout vent length (SVL) with a digital caliper to the nearest 0.1 mm (range: 38.9–56.0; mean ± s.d.: 48.19 ± 5.22 mm). Lizards were individually marked using a cautery pen [77]. For logistical reasons, the first 20 lizards captured were transported to the Station d'Ecologie Théorique et Expérimentale du CNRS à Moulis as the low-elevation treatment (42°57′26.8″ N, 1° 05′08.3″ E; 436 m ASL; $PO_2 \approx 20.1$ kPa) and the following 20 others to the Observatoire du Pic de Midi de Bigorre (42°56′11.09″ N, 0°8′32.9″ E; 2877 m ASL; $PO_2 \approx 15.3$ kPa). These differences in elevation result in about a 25% reduction in oxygen availability at the high elevation Pic du Midi lab compared with the low elevation lab [78]. Treatment groups did not differ in SVL ($t_{37.6} = 0.035$, $p = 0.97$) or mass ($t_{37.9} = 0.071$, $p = 0.94$) at the beginning of the experiment or in mass at the end of the experiment ($t_{36.9} = 1.51$, $p = 0.14$).

Lizards were maintained under identical conditions in both labs so that the primary difference in environments was total atmospheric pressure and thus oxygen availability. While such experimental designs are not able to completely isolate the effects of reduction in oxygen availability from changes in total atmospheric pressure, they are essential complements to experiments that manipulate oxygen concentration in a controlled laboratory setting [79]. Lizards were housed in groups of 2–4 in plastic enclosures (38 × 26 × 23 cm) containing a thin layer of substrate, a water container, and two plastic hides also used as thermoregulation platforms (15 × 5 × 3.5 cm). Every second day, lizards were fed with mealworms (*Tenebrio* sp. larvae) and white maggots (*Calliphora vomitoria*), and water was provided ad libitum. The cages were misted once a day. A UV lamp provided light for 11 h per day, and the enclosures were heated with incandescent heat lamps (42 W) for 6 h per day at 1 h intervals, providing a gradient of 20 to 36 °C. Animals stayed in captivity for 2 to 3 days before the start of the testing schedule (see below).

2.3. Thermal Preferences

We quantified lizard thermal preferences for both treatment groups using standard procedures in a thermal gradient. After two hours acclimating to ambient temperatures (20 °C) in their home cages, four lizards were placed in individual lanes of a thermal preference arena (each lane 90 × 15 cm). On one side of this arena, we suspended four ceramic lamps (150 W) to create a thermal gradient ranging from 20 to 60 °C. The animals were left undisturbed for a one-hour acclimation period. Using two thermal cameras (model C3, Flir Systems, Wilsonville, OR, USA) placed on tripods above the arena at a distance of approximately 1 m, we captured images of the lizards on the gradient every 5 min for 3 h. Temperature data were extracted from the thermographs with Flir Tools (v.6.4, Flir Systems). We extracted data from the image that had the best angle or clearest image of the lizard at a given timepoint. We used the *area box tool* to select pixels in the centre of each lizard dorsum. We used an emissivity value of 0.97, appropriate for reptile skin [80,81], and accounted for ambient temperature and the distance between camera and animal. Using the 36 values for each individual, we quantified the preferred temperature as the mean of the middle 50% (mean of interquartile range [82]). Data for one individual were excluded from analysis because this lizard wedged himself partially under the partition and did not move for the entire trial period.

2.4. Performance Measurements

Our goal was to create thermal performance curves for sprint speed for lizards in both treatment groups. We measured sprint performance at five temperatures, spanning the active range for this species (15, 22, 29, 32, and 35 °C [68,83]). We tested lizards at a maximum of two temperatures per day, with 6–16 h rest between trials. The experiment was conducted over three consecutive days. Before sprinting, we acclimated lizards for one hour in thermostatically controlled incubators (Aqualytic, Germany) at the test temperature. Room temperature was thermostatically controlled to 20 °C. We then sprinted lizards on a

1-m level racetrack lined with artificial turf, recording trials with a video camera (25 fps, Sony Model HDR-XR160E, Sony Corporation, Tokyo, Japan) placed on a tripod directly above the track. Lizards were prompted to sprint from 2 to 6 lengths of the track (i.e., until the visually fastest running speed was obtained). Data were extracted from videos using Tracker software [84]. For each length performed by a lizard, we calculated the maximum speed as the longest distance a lizard traversed in a time step of 0.2 s. We sprinted lizards from both treatments in the same order of test temperatures, from coolest to warmest, so that lizards in both groups would experience the exact same treatment order and to avoid any potentially detrimental effects of incubation at the highest temperatures, especially in the high-elevation treatment where animals may be more sensitive to high temperatures [65]. To account for the potential effects of acclimation on lizard performance, after the trials at the highest temperature, we again sprinted all lizards at 15 °C to test for changes in performance over time in captivity and to estimate repeatability of sprint performance over the duration of the experiment.

2.5. Ethics Statement

Field and lab protocols were conducted under permit from the Direction régionale de l'environnement, de l'aménagement et du logement (DREAL) Midi-Pyrénées (Arrêté Préfectoral No: 2017-s-02 du 30 mars 2017), under current ethical committee approval (APAFIS DAP#16359), and in accordance with Directive 2010/63/EU on protection of animals used for Scientific Purposes. Animals were returned to the site of capture after experiments.

2.6. Statistical Methods

To test for differences between treatment groups in thermal preference, we first used Levene's test to assess differences in variance and a *t*-test, assuming unequal variances, to test for differences in the mean, implemented in the programming language R [85]. We utilised linear mixed models to assess the relative influence of elevation treatment (low/high), temperature (treated as a categorical effect), body size (SVL), and the interaction of treatment and temperature on sprint performance. We $\log_{10}$-transformed sprint speed before analysis to meet the assumption of normal distribution of residuals and included a random intercept for individual to account for repeated measures on the same animal. We used the package *emmeans* for post hoc comparisons of estimated marginal means, corrected for multiple comparisons with the Tukey method, in order to compare sprint speed between treatment groups at each temperature [86,87]. We implemented models with the *lme4* package [88] in R. We confirmed normal distribution of residuals with a Shapiro–Wilk test and determined the relative importance of fixed effects using type III sums of squares, correcting denominator degrees of freedom for *F*-tests [89]. All data figures were created with the *ggplot2* package [90]. Additionally, we assessed whether sprint speed changed over time in captivity by comparing sprint performance at 15 °C between measures made after 3 days in captivity and after 6 days in captivity with a linear mixed model of $\log_{10}$-transformed sprint speed (as above), including the fixed effect of time of measurement and the random effect of individual. Further, we calculated repeatability of sprint performance at this temperature with the *rptR* package in R [91], using 1000 bootstraps and 1000 permutations to estimate 95% confidence intervals and a *p*-value, respectively.

3. Results

Field body temperatures at time of capture averaged 25.0 °C (N = 31; range: 12.6–31.7 °C; Figure 2A). Transplanting lizards to low elevation affected both the mean ($t_{27.6} = -2.92$, $p = 0.0069$) and variance ($F_{1,37} = 14.7$, $p = 0.0005$) of thermal preference: lizards transplanted to low elevation exhibited lower thermal preferences and greater variance compared to lizards maintained at high elevation (low elevation mean ± SD: 30.6 ± 3.6 °C; high elevation mean ± SD: 33.2 ± 1.7 °C; Figure 2B). Elevation treatment, temperature, and the interaction of elevation treatment and temperature affected sprint performance (Table 1, Figure 3). Lizards transplanted to low elevation were slower runners at the three highest temperatures (29, 32, and 35 °C), as demonstrated by post hoc comparison of estimated marginal means (Table 2, Figure 3). Snout-vent length also exhibited a marginal effect on sprint performance, with larger lizards sprinting faster (β = 0.0032 ± 0.0016 SE; Table 1). Sprint speed at 15 °C did not differ between the two measures made at the beginning and end of the experiment (β = −0.023 ± 0.015 SE; $F_{1,39}$ = 2.21, p = 0.14), and individuals exhibited moderate repeatability of sprint performance at this temperature (R = 0.381, 95% CI: 0.084–0.619, p = 0.003).

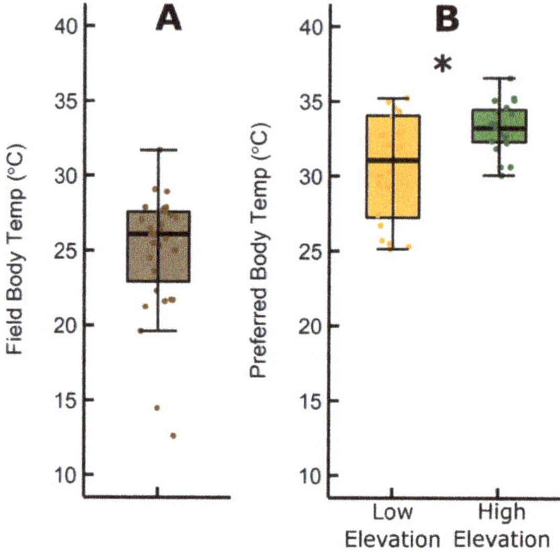

Figure 2. Boxplots and raw values of (**A**) field body temperatures and (**B**) thermal preferences of adult male *Iberolacerta bonnali* lizards. Thermal preferences (panel **B**) measured in lizards at low and high elevation. Tukey boxplots show median, interquartile range, and 1.5× interquartile range of raw data values. Asterisk indicates significant difference between treatment groups for thermal preference (see text for statistical details).

Table 1. Results of linear mixed model analysis of sprint performance (log_{10}-transformed m/s) in adult male *Iberolacerta bonnali* lizards at low and high elevation (see text for statistical details).

Source of Variation	Test Statistics
Temperature	
F (df_n, df_d)	114.8 (4, 152)
Pr > F	**<0.001**
Treatment	
F (df_n, df_d)	5.13 (1, 37)
Pr > F	**0.030**
Temperature × Treatment	
F (df_n, df_d)	2.84 (4, 152)
Pr > F	**0.026**
Snout-vent length	
F (df_n, df_d)	4.02 (1, 37)
Pr > F	0.052

Significant effects shown in bold (p < 0.05).

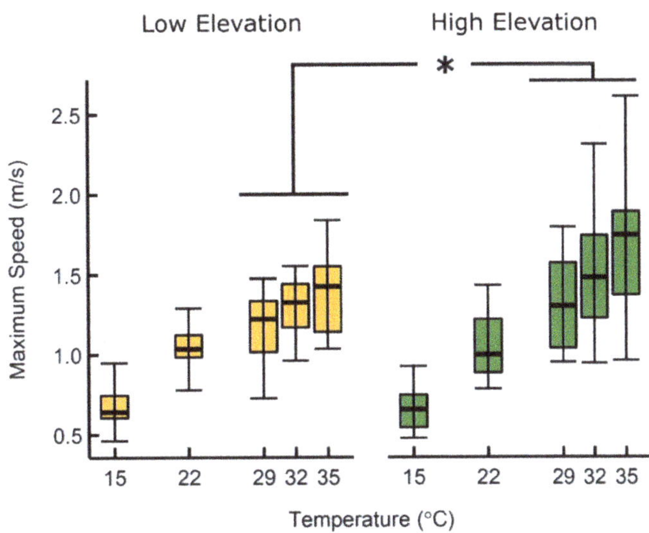

Figure 3. Boxplots of raw values for sprint performance in adult male *Iberolacerta bonnali* lizards at low and high elevation. Asterisk indicates significant difference between treatment groups at 29, 32, and 35 °C (see text for statistical details).

Table 2. Estimated marginal means of $\log_{10}$-transformed transformed sprint speed (m/s) and differences between treatment groups at each temperature in adult male *Iberolacerta bonnali* lizards (see text for statistical details).

Temperature (°C)	Low Elevation (SE)	High Elevation (SE)	Difference (SE)	Significance Test
15	−0.176 (0.02)	−0.188 (0.02)	0.012 (0.29)	$t_{171} = 0.43$ $p = 0.668$
22	0.022 (0.02)	0.013 (0.02)	0.009 (0.29)	$t_{171} = 0.33$ $p = 0.744$
29	0.054 (0.02)	0.113 (0.02)	−0.059 (0.29)	$t_{171} = -2.068$ $\mathbf{p = 0.040}$
32	0.103 (0.02)	0.171 (0.02)	−0.068 (0.29)	$t_{171} = 0.-2.39$ $\mathbf{p = 0.018}$
35	0.134 (0.02)	0.212 (0.02)	−0.079 (0.29)	$t_{171} = -2.77$ $\mathbf{p = 0.006}$

Significant differences at a given temperature shown in bold ($p < 0.05$).

4. Discussion

Previous work on vertebrate ectotherms demonstrates a near-universal limitation of performance and aerobic metabolic capacity under conditions of reduced oxygen availability, especially at high temperatures [65,92]. On the basis of this work, one would predict that a lizard species endemic to high-elevation habitats, and thus reduced oxygen availability, would demonstrate increased performance when exposed to a relatively hyperoxic environment. Our results are exactly contrary to this prediction, but in support of the hyperoxia-as-constraint hypothesis proposed in the introduction to this paper. When transplanted to low elevation, individuals of the high-elevation specialist *I. bonnali* suffered reduced sprint speed at high temperatures (Tables 1 and 2, Figure 3) and selected lower body temperatures in thermal preference trials (Figure 2B). Most of our understanding of oxygen limitation in ectothermic vertebrates comes from experiments reducing oxygen availability relative to the conditions in which organisms evolved or developed, and thus experiments such as the current study are essential for understanding more broadly how organisms can deal with ecologically relevant levels of oxygen variation. Our results suggest strongly that adaptation to reduced oxygen environments restricts an organism's ability to take advantage of increased oxygen availability and, in fact, acute exposure to hyperoxia may be detrimental to performance and ultimately fitness. For example, limiting sprinting performance may have immediate consequences on lizards' ability to avoid predators or catch prey [93,94].

While essential for aerobic life, oxygen is also toxic due to its ability to form molecules which attract electrons and can damage important biochemical structures [95–97]. Our results demonstrate that, even across ecologically relevant levels, acute increase in oxygen availability does not necessarily benefit whole-organism performance. This is in accord with previous work demonstrating that organisms that have evolved at near-sea-level conditions are unable to increase performance measures under hyperoxia [98]. For example, other lizard species exposed to hyperoxia do not alter selected body temperatures or behavioural response to high temperatures, although hyperoxia may increase physiological tolerance to high temperatures ([99–101], but see [102]). This suggests that multiple physiological pathways involved in aerobic respiration are fine-tuned to current oxygen environments, not simply limited by ambient oxygen levels. Therefore, deviations from baseline availability—either increase or decrease—may disrupt these pathways and lead to performance decrements. Future work is needed to examine the specific pathways and trade-offs involved. Our results suggest that, at their native elevation, lizards have evolved to meet metabolic demand when exposed to high temperatures despite low partial pressure of oxygen [103]. When more oxygen is available, these pathways can be dysregulated by hyperoxia in a manner that may disrupt the regulation of oxidative phosphorylation (ATP production) and increase the production of damaging reactive oxygen molecules (ROMs) in the mitochondria (reviewed in [62,63]). In our experiment, lizards transplanted to low elevation lowered their preferred body temperatures, which will result in reduced aerobic metabolic rates. This reduction of oxygen demand in the presence of increased oxygen availability could further exacerbate a potential increase in ROM production [104,105]. At the same time, this reduction in preferred body temperature coincides with a decrement in running performance at high temperatures in hyperoxia. Lizards may be avoiding warmer temperatures where performance is inhibited as a compensatory mechanism. Further work is needed to discern the specific signalling pathways that determine an individual's preferred temperature and how these may be disrupted by relative hyperoxia.

The mechanisms that restrict certain taxa to high-elevation habitats remain elusive. Most commonly, restriction to high-elevation habitats is attributed to either inability to inhabit warmer, low-elevation habitats (e.g., [25]) or due to the presence of competitors at lower elevations [32,106,107], including specific examples in high-elevation specialist lizards of the genus *Iberolacerta* [39,40,42,43,45,108]. However, the congener *Iberolacerta cyreni* does not exhibit agonistic interactions with the low-elevation lizard *Podarcis muralis* either in experimental or field settings, providing little support for competitive exclusion restricting *I. cyreni* to high-elevation habitats [46,109]. Our results suggest that in addition to competitive interactions and cold-specialised thermal physiology, a third mechanism—adaptation to low-oxygen environments and inability to deal with relative hyperoxia—may contribute to the elevation restriction observed in *I. bonnali* and potentially other *Iberolacerta* species. *I. bonnali* are extremely adept thermoregulators [83], suggesting that they could behaviourally buffer themselves in a warmer environment [38,110]. However, levels of oxygen ability will interact with available temperatures to shape thermal ecology (as described by the hierarchical mechanisms of thermal limitation hypothesis [65]). If thermoregulatory set-points are determined by the thermodynamic effects of temperature on metabolism, increased oxygen availability could disrupt the acquisition of optimal temperatures for different aspects of organismal function. For example, we found that *I. bonnali* transplanted to low elevations selected temperatures 2.6 °C cooler than lizards kept close to the elevation of origin. This reduction in selected temperatures will likely result in a decrement in fitness-related physiological processes. Additionally, exposure to novel conditions might increase among-individual variation in traits when previously cryptic genetic variation is exposed [111,112]. In this case, increased among-individual variation in thermal preferences could indicate there is greater genetic variance—in trait means or plasticity—upon which natural selection could act in novel environments.

Our results suggest that the evolved capacities of *I. bonnali* to compensate for low oxygen availability in their high-elevation habitats may be maladaptive when lizards are translocated to low elevations and increased oxygen availability. The extent to which this limitation may be important in restricting the range of high-elevation specialists needs to be assessed in more taxa. The inability to adjust to relative hyperoxia may act in conjunction with other factors, such as increased interspecific competition or higher temperatures, to limit species distributions. Our data also suggest that *I. bonnali* may be resilient to short-term increases in high temperatures. Their preferred body temperature in unconstrained laboratory conditions is well above temperatures lizards achieved in the field and they are capable of maintaining at least one measure of whole-organism performance, sprinting, at even the highest temperature we tested (35 °C). The important conservation question is then the capacity for lower-elevation species, such as lizards in the genus *Podarcis*, to move upslope and the potential ramifications of increased interspecific interactions, which remain unclear [45,46,113].

Future studies should also test the response of high-elevation lizards when acclimated to low-elevation conditions for longer periods of time and how physiological plasticity might mitigate the negative consequences we observed, such as through shifts in blood oxygen capacity, reactive oxygen molecule production, or metabolic rates (e.g., [114]). Over longer exposures to relative hyperoxia, lizards may be able to respond via physiological plasticity to compensate for the new environment. For example, the congener *I. cyreni* dramatically reduced hematocrit, increased body condition, and increased preferred body temperatures after two weeks of exposure to a modest increase in oxygen availability [114]. However, such plasticity may not fully compensate, and performance can be reduced, as found in low-elevation lizards transplanted to high elevation (e.g., [66,115]). It is also essential to test the capacity of embryos to develop successfully in different oxygen environments, as this life-history stage may be more resilient to such limitations [48,116–118]. Our results were directly opposite to predictions based on models developed from studies of organisms inhabiting generally normoxic environments introduced to conditions of oxygen limitation. This highlights the complexities of oxygen physiology and that the assumption of "more is better" does not apply to organisms adapted to life at high elevation. Studies of the unique physiological adaptations of high-elevation organisms remain an essential—and underexplored—area in characterising the vast biological diversity of our planet [119,120].

Author Contributions: Conceptualisation, E.J.G. and F.A.; data curation, S.S., L.K. and F.A.; formal analysis, E.J.G.; funding acquisition, F.A.; investigation, S.S., L.K. and F.A.; methodology, E.J.G., S.S., C.P. and F.A.; project administration, F.A.; resources, F.A.; supervision, E.J.G. and F.A.; writing—original draft, E.J.G., L.K. and F.A.; writing—review and editing, E.J.G., S.S., L.K., C.P. and F.A. All authors have read and agreed to the published version of the manuscript.

Funding: The Summer Science Research Program at Ohio Wesleyan University, coordinated by L. Tuhela-Reuning, providing funding and support for Sierra Spears. We thank European INTERREG POCTEFA OPCC ADAPYR no. EFA346/19 and TULIP Laboratory of Excellence (ANR-10-LABX-41) for funding.

Data Availability Statement: All data used in analyses are freely available in the Mendeley Data Repository (http://dx.doi.org/10.17632/pdwgvtc6sn.1, accessed on 1 April 2019).

Acknowledgments: We are grateful to the staff of the Observatoire Midi-Pyrénées at Pic du Midi for their hospitality. We also thank B. Bodensteiner for technical advice and support.

Conflicts of Interest: The authors declare no conflict of interest.

References

1. Sayre, R.; Frye, C.; Karagulle, D.; Krauer, J.; Breyer, S.; Aniello, P.; Wright, D.J.; Payne, D.; Adler, C.; Warner, H.; et al. A New High-Resolution Map of World Mountains and an Online Tool for Visualizing and Comparing Characterizations of Global Mountain Distributions. *Mt. Res. Dev.* **2018**, *38*, 240–249. [CrossRef]
2. Hoorn, C.; Mosbrugger, V.; Mulch, A.; Antonelli, A. Biodiversity from mountain building. *Nat. Geosci.* **2013**, *6*, 154. [CrossRef]
3. Shen, C.; Shi, Y.; Fan, K.; He, J.-S.; Adams, J.M.; Ge, Y.; Chu, H. Soil pH dominates elevational diversity pattern for bacteria in high elevation alkaline soils on the Tibetan Plateau. *FEMS Microbiol. Ecol.* **2019**, *95*, 95. [CrossRef] [PubMed]
4. King, A.J.; Freeman, K.R.; McCormick, K.F.; Lynch, R.C.; Lozupone, C.; Knight, R.; Schmidt, S.K. Biogeography and habitat modelling of high-alpine bacteria. *Nat. Commun.* **2010**, *1*, 53. [CrossRef] [PubMed]
5. Birrell, J.H.; Shah, A.A.; Hotaling, S.; Giersch, J.J.; Williamson, C.E.; Jacobsen, D.; Woods, H.A. Insects in high-elevation streams: Life in extreme environments imperiled by climate change. *Glob. Chang. Biol.* **2020**, *26*, 6667–6684. [CrossRef]
6. Mani, M.S. *Ecology and Biogeography of High Altitude Insects*; Springer: Berlin/Heidelberg, Germany, 2013; Volume 4.
7. Thaler, K. The diversity of high altitude arachnids (Araneae, Opiliones, Pseudoscorpiones) in the Alps. In *Alpine Biodiversity in Europe*; Springer: Berlin/Heidelberg, Germany, 2003; pp. 281–296.
8. Baur, B.; Meier, T.; Baur, A.; Schmera, D. Terrestrial gastropod diversity in an alpine region: Disentangling effects of elevation, area, geometric constraints, habitat type and land-use intensity. *Ecography* **2014**, *37*, 390–401. [CrossRef]
9. Ports, M. Habitat affinities and distributions of land gastropods from the Ruby Mountains and East Humboldt Range of northeastern Nevada. *Veliger* **1996**, *39*, 335–341.
10. Atkinson, C.L.; Alexiades, A.V.; MacNeill, K.L.; Encalada, A.C.; Thomas, S.A.; Flecker, A.S. Nutrient recycling by insect and fish communities in high-elevation tropical streams. *Hydrobiologia* **2019**, *838*, 13–28. [CrossRef]
11. Tong, C.; Fei, T.; Zhang, C.; Zhao, K. Comprehensive transcriptomic analysis of Tibetan Schizothoracinae fish *Gymnocypris przewalskii* reveals how it adapts to a high altitude aquatic life. *BMC Evol. Biol.* **2017**, *17*, 74. [CrossRef]
12. Cordier, J.M.; Lescano, J.N.; Ríos, N.E.; Leynaud, G.C.; Nori, J. Climate change threatens micro-endemic amphibians of an important South American high-altitude center of endemism. *Amphib. Reptil.* **2020**, *41*, 233–243. [CrossRef]
13. Yang, X.; Wang, Y.; Zhang, Y.; Lee, W.-H.; Zhang, Y. Rich diversity and potency of skin antioxidant peptides revealed a novel molecular basis for high-altitude adaptation of amphibians. *Sci. Rep.* **2016**, *6*, 19866. [CrossRef]
14. Storz, J.F. Hemoglobin function and physiological adaptation to hypoxia in high-altitude mammals. *J. Mammal.* **2007**, *88*, 24–31. [CrossRef]
15. Badgley, C. Tectonics, topography, and mammalian diversity. *Ecography* **2010**, *33*, 220–231. [CrossRef]
16. Fjeldså, J.; Bowie, R.; Rahbek, C. The role of mountain ranges in the diversification of birds. *Annu. Rev. Ecol. Evol. Syst.* **2012**, *43*, 249–265. [CrossRef]
17. Cerdeña, J.; Farfán, J.; Quiroz, A.J. A high mountain lizard from Peru: The world's highest-altitude reptile. *Herpetozoa* **2021**, *34*, 61. [CrossRef]
18. McCain, C.M. Global analysis of reptile elevational diversity. *Glob. Ecol. Biogeogr.* **2010**, *19*, 541–553. [CrossRef]
19. Cadena, C.D.; Kozak, K.H.; Gómez, J.P.; Parra, J.L.; McCain, C.M.; Bowie, R.C.K.; Carnaval, A.C.; Moritz, C.; Rahbek, C.; Roberts, T.E.; et al. Latitude, elevational climatic zonation and speciation in New World vertebrates. *Proc. R. Soc. B* **2012**, *279*, 194–201. [CrossRef]
20. Hodkinson, I.D. Terrestrial insects along elevation gradients: Species and community responses to altitude. *Biol. Rev.* **2005**, *80*, 489–513. [CrossRef]
21. Kindler, C.; Böhme, W.; Corti, C.; Gvoždík, V.; Jablonski, D.; Jandzik, D.; Metallinou, M.; Široký, P.; Fritz, U. Mitochondrial phylogeography, contact zones and taxonomy of grass snakes (*Natrix natrix, N. megalocephala*). *Zool. Scr.* **2013**, *42*, 458–472. [CrossRef]
22. Marshall, L.; Perdijk, F.; Dendoncker, N.; Kunin, W.; Roberts, S.; Biesmeijer, J.C. Bumblebees moving up: Shifts in elevation ranges in the Pyrenees over 115 years. *Proc. R. Soc. B* **2020**, *287*, 20202201. [CrossRef]
23. Dupoué, A.; Trochet, A.; Richard, M.; Sorlin, M.; Guillon, M.; Teuliere-Quillet, J.; Vallé, C.; Rault, C.; Berroneau, M.; Berroneau, M.; et al. Genetic and demographic trends from rear to leading edge are explained by climate and forest cover in a cold-adapted ectotherm. *Divers. Distrib.* **2020**. [CrossRef]
24. Gifford, M.E.; Kozak, K.H. Islands in the sky or squeezed at the top? Ecological causes of elevational range limits in montane salamanders. *Ecography* **2012**, *35*, 193–203. [CrossRef]
25. Wiens, J.J.; Camacho, A.; Goldberg, A.; Jezkova, T.; Kaplan, M.E.; Lambert, S.M.; Miller, E.C.; Streicher, J.W.; Walls, R.L. Climate-change, extinction, and Sky Island biogeography in a montane lizard. *Mol. Ecol.* **2019**, *28*, 2610–2624. [CrossRef]
26. Monge, C.; Leon-Velarde, F. Physiological adaptation to high altitude: Oxygen transport in mammals and birds. *Physiol. Rev.* **1991**, *71*, 1135–1172. [CrossRef] [PubMed]
27. González-Morales, J.C.; Beamonte-Barrientos, R.; Bastiaans, E.; Guevara-Fiore, P.; Quintana, E.; Fajardo, V. A mountain or a plateau? Hematological traits vary nonlinearly with altitude in a highland lizard. *Physiol. Biochem. Zool.* **2017**, *90*, 638–645. [CrossRef] [PubMed]
28. Zhao, X.; Wu, N.; Zhu, Q.; Gaur, U.; Gu, T.; Li, D. High-altitude adaptation of Tibetan chicken from MT-COI and ATP-6 perspective. *Mitochondrial DNA A* **2016**, *27*, 3280–3288. [CrossRef]

29. Sinervo, B.; Méndez-De-La-Cruz, F.; Miles, D.B.; Heulin, B.; Bastiaans, E.; Cruz, M.V.-S.; A Lararesendiz, R.; Martínez-Méndez, N.; Calderón-Espinosa, M.L.; Meza-Lázaro, R.N.; et al. Erosion of Lizard Diversity by Climate Change and Altered Thermal. *Niches Sci.* **2010**, *328*, 894–899. [CrossRef]
30. Rozen-Rechels, D.; Dupoué, A.; Lourdais, O.; Chamaillé-Jammes, S.; Meylan, S.; Clobert, J.; Le Galliard, J. When water interacts with temperature: Ecological and evolutionary implications of thermo-hydroregulation in terrestrial ectotherms. *Ecol. Evol.* **2019**, *9*, 10029–10043. [CrossRef] [PubMed]
31. Bachmann, J.C.; Van Buskirk, J. Adaptation to elevation but limited local adaptation in an amphibian. *Evolution* **2020**. [CrossRef]
32. Brown, J.H.; Stevens, G.C.; Kaufman, D.M. The geographic range: Size, shape, boundaries, and internal structure. *Annu. Rev. Ecol. Syst.* **1996**, *27*, 597–623. [CrossRef]
33. Tingley, R.; Vallinoto, M.; Sequeira, F.; Kearney, M.R. Realized niche shift during a global biological invasion. *Proc. Natl. Acad. Sci. USA* **2014**, *111*, 10233–10238. [CrossRef] [PubMed]
34. Alexander, J.M.; Diez, J.M.; Levine, J.M. Novel competitors shape species' responses to climate change. *Nature* **2015**, *525*, 515–518. [CrossRef]
35. Pottier, G. *Plan National d'Actions en Faveur des Lézards des Pyrénées*; Nature Midi-Pyrénées: Bagnères de Bigorre, France, 2012; p. 125.
36. Berroneau, M. *Atlas des Amphibiens et Reptiles d'Aquitaine*; Cistude Nature: Le Haillan, Gironde, 2014; 257p.
37. Datcharry, R. *Détectabilité en Montagne de Deux Lézards Rupicoles: Le Lézard des Murailles (Podarcis muralis) vs. le Lézard de Bonnal (Iberolacerta bonnali). Mémoire de Stage de Master 2 GBAT*; Université Paul Sabatier: Toulouse, France, 2014.
38. Aguado, S.; Braña, F. Thermoregulation in a cold-adapted species (Cyren's Rock Lizard, *Iberolacerta cyreni*): Influence of thermal environment and associated costs. *Can. J. Zool.* **2014**, *92*, 955–964. [CrossRef]
39. Monasterio, C.; Shoo, L.P.; Salvador, A.; Siliceo, I.; Díaz, J.A. Thermal constraints on embryonic development as a proximate cause for elevational range limits in two Mediterranean lacertid lizards. *Ecography* **2011**, *34*, 1030–1039. [CrossRef]
40. Monasterio, C.; Verdú-Ricoy, J.; Salvador, A.; Díaz, J.A. Living at the edge: Lower success of eggs and hatchlings at lower elevation may shape range limits in an alpine lizard. *Biol. J. Linn. Soc.* **2016**, *118*, 829–841. [CrossRef]
41. Osojnik, N.; Žagar, A.; Carretero, M.A.; García-Muñoz, E.; Vrezec, A. Ecophysiological dissimilarities of two sympatric lizards. *Herpetologica* **2013**, *69*, 445–454. [CrossRef]
42. Žagar, A.; Carretero, M.A.; Marguč, D.; Simčič, T.; Vrezec, A. A metabolic syndrome in terrestrial ectotherms with different elevational and distribution patterns. *Ecography* **2018**, *41*, 1728–1739. [CrossRef]
43. Carranza, S.; Arnold, E.N.; Amat, F. DNA phylogeny of Lacerta (*Iberolacerta*) and other lacertine lizards (Reptilia: Lacertidae): Did competition cause long-term mountain restriction? *Syst. Biodivers.* **2004**, *2*, 57–77. [CrossRef]
44. Žagar, A.; Carretero, M.A.; Osojnik, N.; Sillero, N.; Vrezec, A. A place in the sun: Interspecific interference affects thermoregulation in coexisting lizards. *Behav. Ecol. Sociobiol.* **2015**, *69*, 1127–1137. [CrossRef]
45. Žagar, A.; Carretero, M.A.; Vrezec, A.; Drašler, K.; Kaliontzopoulou, A. Towards a functional understanding of species coexistence: Ecomorphological variation in relation to whole-organism performance in two sympatric lizards. *Funct. Ecol.* **2017**, *31*, 1780–1791. [CrossRef]
46. Monasterio, C.; Salvador, A.; Diaz, J.A. Competition with wall lizards does not explain the alpine confinement of Iberian rock lizards: An experimental approach. *Zoology* **2010**, *113*, 275–282. [CrossRef]
47. Cordero, G.A.; Karnatz, M.L.; Svendsen, J.C.; Gangloff, E.J. Effects of low-oxygen conditions on embryo growth in the painted turtle. *Chrysemys. Picta. Integr. Zool.* **2017**, *12*, 148–156. [CrossRef]
48. Kouyoumdjian, L.; Gangloff, E.J.; Souchet, J.; Cordero, G.A.; Dupoue, A.; Aubret, F. Transplanting gravid lizards to high elevation alters maternal and embryonic oxygen physiology, but not reproductive success or hatchling phenotype. *J. Exp. Biol.* **2019**, *222*, jeb206839. [CrossRef]
49. Mallard, F. *Programme les Sentinelles du Climat–Tome IX: Connaitre et Comprendre pour Protéger les Espèces Animales et Végétales face au Changement Climatique*; Cistude Nature: Le Haillan, Gironde, 2020; 822p.
50. Pecl, G.T.; Araújo, M.B.; Bell, J.D.; Blanchard, J.; Bonebrake, T.C.; Chen, I.-C.; Clark, T.D.; Colwell, R.K.; Danielsen, F.; Evengård, B.; et al. Biodiversity redistribution under climate change: Impacts on ecosystems and human well-being. *Science* **2017**, *355*, eaai9214. [CrossRef]
51. Parmesan, C. Ecological and evolutionary responses to recent climate change. *Annu. Rev. Ecol. Evol. Syst.* **2006**, *37*, 637–669. [CrossRef]
52. Chen, I.-C.; Hill, J.K.; Ohlemüller, R.; Roy, D.B.; Thomas, C.D. Rapid range shifts of species associated with high levels of climate warming. *Science* **2011**, *333*, 1024–1026. [CrossRef]
53. Bravo, D.N.; Araújo, M.B.; Lasanta, T.; Moreno, J.I.L. Climate change in Mediterranean mountains during the 21st century. *AMBIO J. Hum. Environ.* **2008**, *37*, 280–285. [CrossRef]
54. Walters, R.J.; Blanckenhorn, W.U.; Berger, D. Forecasting extinction risk of ectotherms under climate warming: An evolutionary perspective. *Funct. Ecol.* **2012**, *26*, 1324–1338. [CrossRef]
55. Beninde, J.; Feldmeier, S.; Veith, M.; Hochkirch, A. Admixture of hybrid swarms of native and introduced lizards in cities is determined by the cityscape structure and invasion history. *Proc. R. Soc. B* **2018**, *285*, 20180143. [CrossRef]
56. Brown, R.M.; Gist, D.H.; Taylor, D.H. Home range ecology of an introduced population of the European wall lizard *Podarcis muralis* (Lacertilia; Lacertidae) in Cincinnati, Ohio. *Am. Midl. Nat.* **1995**, *133*, 344–359. [CrossRef]

57. While, G.M.; Williamson, J.; Prescott, G.; Horvathova, T.; Fresnillo, B.; Beeton, N.J.; Halliwell, B.; Michaelides, S.; Uller, T. Adaptive responses to cool climate promotes persistence of a non-native lizard. *Proc. R. Soc. B* **2015**, *282*, 20142638. [CrossRef] [PubMed]
58. Hedeen, S.E.; Hedeen, D.L. Railway-aided dispersal of an introduced *Podarcis muralis* population. *Herp. Rev.* **1999**, *30*, 57.
59. Schaefer, J.; Walters, A. Metabolic cold adaptation and developmental plasticity in metabolic rates among species in the *Fundulus notatus* species complex. *Funct. Ecol.* **2010**, *24*, 1087–1094. [CrossRef]
60. Chown, S.L.; Gaston, K.J. Exploring links between physiology and ecology at macro-scales: The role of respiratory metabolism in insects. *Biol. Rev.* **1999**, *74*, 87–120. [CrossRef]
61. Lardies, M.A.; Bacigalupe, L.D.; Bozinovic, F. Testing the metabolic cold adaptation hypothesis: An intraspecific latitudinal comparison in a common woodlouse. *Evol. Ecol. Res.* **2004**, *6*, 567–578.
62. Muller, F.L.; Lustgarten, M.S.; Jang, Y.; Richardson, A.; Van Remmen, H. Trends in oxidative aging theories. *Free Radic. Biol. Med.* **2007**, *43*, 477–503. [CrossRef]
63. Costantini, D. Oxidative stress ecology and the d-ROMs test: Facts, misfacts and an appraisal of a decade's work. *Behav. Ecol. Sociobiol.* **2016**, *70*, 809–820. [CrossRef]
64. Balaban, R.S.; Nemoto, S.; Finkel, T. Mitochondria, oxidants, and aging. *Cell* **2005**, *120*, 483–495. [CrossRef] [PubMed]
65. Gangloff, E.J.; Telemeco, R.S. High temperature, oxygen, and performance: Insights from reptiles and amphibians. *Integr. Comp. Biol.* **2018**, *58*, 9–24. [CrossRef] [PubMed]
66. Gangloff, E.J.; Sorlin, M.; Cordero, G.A.; Souchet, J.; Aubret, F. Lizards at the peak: Physiological plasticity does not maintain performance in lizards transplanted to high altitude. *Physiol. Biochem. Zool.* **2019**, *92*, 189–200. [CrossRef]
67. Li, X.; Wu, P.; Ma, L.; Huebner, C.; Sun, B.; Li, S. Embryonic and post-embryonic responses to high-elevation hypoxia in a low-elevation lizard. *Integr. Zool.* **2020**, *15*, 338–348. [CrossRef]
68. Arribas, O.J. Habitat selection, thermoregulation and activity of the Pyrenean Rock Lizard *Iberolacerta bonnali* (Lantz, 1927). *Herpetozoa* **2009**, *22*, 146–166.
69. Arribas, O.J. Lagartija pirenaica–*Iberolacerta bonnali*. In *Enciclopedia Virtual de los Vertebrados Españoles*; Salvador, A., Marco, A., Eds.; Museo Nacional de Ciencias Naturales: Madrid, Spain, 2015.
70. Galán, P.; Arribas, O. Reproductive characteristics of the Pyrenean high-mountain lizards: *Iberolacerta aranica* (Arribas, 1993), *I. aurelioi* (Arribas, 1994) and *I. bonnali* (Lantz, 1927). *Anim. Biol.* **2005**, *55*, 163–190. [CrossRef]
71. Pottier, G.; Arthur, C.-P.; Weber, L.; Cheylan, M. Répartition des lézards du genre *Iberolacerta* Arribas, 1997 (Sauria: Lacertidae) en France. 3/3: Le Lézard de Bonnal, *Iberolacerta bonnali* (Lantz, 1927). *Bull. Soc. Herp. Fr.* **2013**, *148*, 425–450.
72. Sinervo, B.; Hedges, R.; Adolph, S.C. Decreased sprint speed as a cost of reproduction in the lizard *Sceloporus occidentalis*: Variation among populations. *J. Exp. Biol.* **1991**, *155*, 323–336. [CrossRef]
73. Díaz de la Vega-Pérez, A.H.; Barrios-Montiel, R.; Jiménez-Arcos, V.H.; Bautista, A.; Bastiaans, E. High-mountain altitudinal gradient influences thermal ecology of the Mesquite Lizard (*Sceloporus grammicus*). *Can. J. Zool.* **2019**, *97*, 659–668. [CrossRef]
74. Mathies, T.; Andrews, R. Influence of pregnancy on the thermal biology of the lizard, *Sceloporus jarrovi*: Why do pregnant females exhibit low body temperatures? *Funct. Ecol.* **1997**, *11*, 498–507. [CrossRef]
75. Blomberg, S.; Shine, R. Reptiles. In *Ecological Census Techniques: A Handbook*, 2nd ed.; Sutherland, W.J., Ed.; Cambridge University Press: Cambridge, UK, 2006; pp. 297–307.
76. Fitzgerald, L.A. Finding and Capturing Reptiles. In *Reptile Biodiversity: Standard Methods for Inventory and Monitoring*; McDiarmid, R.W., Foster, M.S., Guyer, C., Chernoff, N., Gibbons, J.W., Eds.; University of California Press: Oakland, CA, USA, 2012; pp. 77–88.
77. Vervust, B.; Van Damme, R. Marking lizards by heat branding. *Herp. Rev.* **2009**, *40*, 173.
78. Bouverot, P. *Adaptation to Altitude-Hypoxia in Vertebrates*; Spring: Berlin, Germany, 1985; p. 176.
79. Ewalts, M.; Dawkins, T.; Friend, A.T. Hypoxia research: To control or not to control? That is the question. *J. Physiol.* **2021**, *599*, 2141–2142. [CrossRef] [PubMed]
80. Luna, S.; Font, E. Use of an infrared thermographic camera to measure field body temperatures of small lacertid lizards. *Herp. Rev.* **2013**, *44*, 59–62.
81. Tattersall, G.J. Infrared thermography: A non-invasive window into thermal physiology. *Comp. Biochem. Physiol. A* **2016**, *202*, 78–98. [CrossRef] [PubMed]
82. Taylor, E.N.; Diele-Viegas, L.M.; Gangloff, E.J.; Hall, J.M.; Halpern, B.; Massey, M.D.; Rödder, D.; Rollinson, N.; Spears, S.; Sun, B.J.; et al. The thermal ecology and physiology of reptiles and amphibians: A user's guide. *J. Exp. Zool. A* **2020**. [CrossRef] [PubMed]
83. Ortega, Z.; Mencia, A.; Perez-Mellado, V. The peak of thermoregulation effectiveness: Thermal biology of the Pyrenean rock lizard, *Iberolacerta bonnali* (Squamata, Lacertidae). *J. Therm. Biol.* **2016**, *56*, 77–83. [CrossRef]
84. Brown, D. *Tracker: Video Analysis and Modeling Tool*; Version 5.1.3; 2019; Available online: https://physlets.org/tracker/ (accessed on 1 April 2019).
85. R Core Team. *R: A Language and Environment for Statistical Computing*; R Foundation for Statistical Computing: Vienna, Austria, 2018.
86. Lenth, R.V. Least-squares means: The R package lsmeans. *J. Stat. Soft.* **2016**, *69*, 1–33. [CrossRef]
87. Lenth, R.V. *Emmeans: Estimated Marginal Means, Aka Least-Squares Means*; R Package Version 1.3.3; 2019; Available online: https://CRAN.R-project.org/package=emmeans (accessed on 1 April 2019).

88. Bates, D.M.; Maechler, M.; Bolker, B.M.; Walker, S. Fitting linear mixed-effects models using lme4. *J. Stat. Soft.* **2015**, *67*, 1–48. [CrossRef]
89. Kenward, M.G.; Roger, J.H. Small sample inference for fixed effects from restricted maximum likelihood. *Biometrics* **1997**, *53*, 983–997. [CrossRef]
90. Wickham, H. *Ggplot2: Elegant Graphics for Data Analysis*; Springer: New York, NY, USA, 2009; p. 213.
91. Stoffel, M.A.; Nakagawa, S.; Schielzeth, H.; Goslee, S. RptR: Repeatability estimation and variance decomposition by generalized linear mixed-effects models. *Method Ecol. Evol.* **2017**, *8*, 1639–1644. [CrossRef]
92. Jackson, D.C. Temperature and hypoxia in ectothermic tetrapods. *J. Therm. Biol.* **2007**, *32*, 125–133. [CrossRef]
93. Irschick, D.J.; Garland, T., Jr. Integrating function and ecology in studies of adaptation: Investigations of locomotor capacity as a model system. *Ann. Rev. Ecol. Syst.* **2001**, *32*, 367–396. [CrossRef]
94. Miles, D.B. The race goes to the swift: Fitness consequences of variation in sprint performance in juvenile lizards. *Evol. Ecol. Res.* **2004**, *6*, 63–75.
95. Barja, G. Aging in vertebrates, and the effect of caloric restriction: A mitochondrial free radical production-DNA damage mechanism? *Biol. Rev. Camb. Philos. Soc.* **2004**, *79*, 235–251. [CrossRef] [PubMed]
96. Damiani, E.; Donati, A.; Girardis, M. Oxygen in the critically ill: Friend or foe? *Curr. Opin. Anesthesio.* **2018**, *31*, 129–135. [CrossRef]
97. Finkel, T.; Holbrook, N.J. Oxidants, oxidative stress and the biology of ageing. *Nature* **2000**, *408*, 239–247. [CrossRef] [PubMed]
98. Seibel, B.A.; Deutsch, C. Oxygen supply capacity in animals evolves to meet maximum demand at the current oxygen partial pressure regardless of size or temperature. *J. Exp. Biol.* **2020**, *223*, jeb210492. [CrossRef] [PubMed]
99. Guerrero, A.C.; VandenBrooks, J.M.; Riley, A.; Telemeco, R.S.; Angilletta, M.J., Jr. Oxygen supply did not affect how lizards responded to thermal stress. *Integr. Zool.* **2018**, *13*, 428–436. [CrossRef]
100. Hicks, J.W.; Wood, S.C. Temperature regulation in lizards: Effects of hypoxia. *Am. J. Physiol. Reg. Integr. Comp. Physiol.* **1985**, *248*, R595–R600. [CrossRef]
101. He, J.; Xiu, M.; Tang, X.; Wang, N.; Xin, Y.; Li, W.; Chen, Q. Thermoregulatory and metabolic responses to hypoxia in the oviparous lizard, *Phrynocephalus przewalskii*. *Comp. Biochem. Physiol. A* **2013**, *165*, 207–213. [CrossRef]
102. Shea, T.K.; DuBois, P.M.; Claunch, N.M.; Murphey, N.E.; Rucker, K.A.; Brewster, R.A.; Taylor, E.N. Oxygen concentration affects upper thermal tolerance in a terrestrial vertebrate. *Comp. Biochem. Physiol. A* **2016**, *199*, 87–94. [CrossRef]
103. McClelland, G.B.; Scott, G.R. Evolved mechanisms of aerobic performance and hypoxia resistance in high-altitude natives. *Annu. Rev. Physiol.* **2019**, *81*, 561–583. [CrossRef]
104. Voituron, Y.; Boel, M.; Roussel, D. Mitochondrial threshold for H_2O_2 release in skeletal muscle of mammals. *Mitochondrion* **2020**, *54*, 85–91. [CrossRef]
105. Koch, R.E.; Buchanan, K.L.; Casagrande, S.; Crino, O.; Dowling, D.K.; Hill, G.E.; Hood, W.R.; McKenzie, M.; Mariette, M.M.; Noble, D.W.A.; et al. Integrating mitochondrial aerobic metabolism into ecology and evolution. *Trends Ecol. Evol.* **2021**. [CrossRef] [PubMed]
106. Lomolino, M.V.; Riddle, B.R.; Whittaker, A.L.; Brown, J.H. *Biogeography*; Sinauer Associates: Sunderland, MA, USA, 2010; p. 560.
107. Jankowski, J.E.; Robinson, S.K.; Levey, D.J. Squeezed at the top: Interspecific aggression may constrain elevational ranges in tropical birds. *Ecology* **2010**, *91*, 1877–1884. [CrossRef] [PubMed]
108. Ortega, Z.; Mencia, A.; Perez-Mellado, V. Are mountain habitats becoming more suitable for generalist than cold-adapted lizards thermoregulation? *PeerJ* **2016**, *4*, e2085. [CrossRef] [PubMed]
109. Monasterio, C.; Salvador, A.; Iraeta, P.; Díaz, J.A. The effects of thermal biology and refuge availability on the restricted distribution of an alpine lizard. *J. Biogeogr.* **2009**, *36*, 1673–1684. [CrossRef]
110. Ortega, Z.; Mencía, A.; Pérez-Mellado, V. Behavioral buffering of global warming in a cold-adapted lizard. *Ecol. Evol.* **2016**, *6*, 4582–4590. [CrossRef]
111. Levis, N.A.; Pfennig, D.W. Evaluating 'Plasticity-first' evolution in nature: Key criteria and empirical approaches. *Trends Ecol. Evol.* **2016**, *31*, 563–574. [CrossRef]
112. Lande, R. Adaptation to an extraordinary environment by evolution of phenotypic plasticity and genetic assimilation. *J. Evol. Biol.* **2009**, *22*, 1435–1446. [CrossRef]
113. Žagar, A.; Simčič, T.; Carretero, M.A.; Vrezec, A. The role of metabolism in understanding the altitudinal segregation pattern of two potentially interacting lizards. *Comp. Biochem. Physiol. A* **2015**, *179*, 1–6. [CrossRef]
114. Megía-Palma, R.; Jiménez-Robles, O.; Hernández-Agüero, J.A.; De la Riva, I. Plasticity of haemoglobin concentration and thermoregulation in a mountain lizard. *J. Therm. Biol.* **2020**, *92*, 102656. [CrossRef]
115. Jiang, Z.-W.; Ma, L.; Mi, C.-R.; Du, W.-G. Effects of hypoxia on the thermal physiology of a high-elevation lizard: Implications for upslope-shifting species. *Biol. Lett.* **2021**, *17*, 20200873. [CrossRef]
116. Cordero, G.A.; Andersson, B.A.; Souchet, J.; Micheli, G.; Noble, D.W.A.; Gangloff, E.J.; Uller, T.; Aubret, F. Physiological plasticity in lizard embryos exposed to high-altitude hypoxia. *J. Exp. Zool. A* **2017**, *327*, 423–432. [CrossRef] [PubMed]
117. Souchet, J.; Bossu, C.; Darnet, E.; Le Chevalier, H.; Poignet, M.; Trochet, A.; Bertrand, R.; Calvez, O.; Martinez-Silvestre, A.; Mossoll-Torres, M.; et al. High temperatures limit developmental resilience to high-elevation hypoxia in the snake *Natrix maura* (Squamata: Colubridae). *Biol. J. Linn. Soc.* **2020**, *132*, 116–133. [CrossRef]

118. Souchet, J.; Gangloff, E.J.; Micheli, G.; Bossu, C.; Trochet, A.; Bertrand, R.; Clobert, J.; Calvez, O.; Martinez-Silvestre, A.; Darnet, E.; et al. High-elevation hypoxia impacts perinatal physiology and performance in a potential montane colonizer. *Integr. Zool.* **2020**, *15*, 544–557. [CrossRef] [PubMed]
119. Storz, J.F. High-altitude adaptation: Mechanistic insights from integrated genomics and physiology. *Mol. Biol. Evol.* **2021**. [CrossRef]
120. Li, J.T.; Gao, Y.D.; Xie, L.; Deng, C.; Shi, P.; Guan, M.L.; Huang, S.; Ren, J.L.; Wu, D.D.; Ding, L.; et al. Comparative genomic investigation of high-elevation adaptation in ectothermic snakes. *Proc. Natl. Acad. Sci. USA* **2018**, *115*, 8406–8411. [CrossRef]

MDPI

Article

Inter-Individual Differences in Ornamental Colouration in a Mediterranean Lizard in Relation to Altitude, Season, Sex, Age, and Body Traits

Gregorio Moreno-Rueda [1], Senda Reguera [1], Francisco J. Zamora-Camacho [2] and Mar Comas [1,*]

[1] Departamento de Zoología, Facultad de Ciencias, Universidad de Granada, E-18071 Granada, Spain; gmr@ugr.es (G.M.-R.); sendareguera@gmail.com (S.R.)

[2] Departamento de Biodiversidad, Ecología y Evolución, Facultad de Ciencias Biológicas, Universidad Complutense de Madrid, E-28040 Madrid, Spain; zamcam@ugr.es

* Correspondence: marcomas@ugr.es

Abstract: Animals frequently show complex colour patterns involved in social communication, which attracts great interest in evolutionary and behavioural ecology. Most researchers interpret that each colour in animals with multiple patches may either signal a different bearer's trait or redundantly convey the same information. Colour signals, moreover, may vary geographically and according to bearer qualities. In this study, we analyse different sources of colour variation in the eastern clade of the lizard *Psammodromus algirus*. Sexual dichromatism markedly differs between clades; both possess lateral blue eyespots, but whereas males in the western populations display strikingly colourful orange-red throats during the breeding season, eastern lizards only show some commissure pigmentation and light yellow throats. We analyse how different colour traits (commissure and throat colouration, and the number of blue eyespots) vary according to body size, head size (an indicator of fighting ability), and sex along an elevational gradient. Our findings show that blue eyespots function independently from colour patches in the commissure and throat, which were interrelated. Males had more eyespots and orange commissures (which were yellow or colourless in females). Throat colour saturation and the presence of coloured commissures increased in older lizards. The number of eyespots, presence of a coloured commissure, and throat colour saturation positively related to head size. However, while the number of eyespots was maximal at lowlands, throat colour saturation increased with altitude. Overall, our results suggest that this lizard harbours several colour signals, which altitudinally differ in their importance, but generally provide redundant information. The relevance of each signal may depend on the context. For example, all signals indicate head size, but commissure colouration may work well at a short distance and when the lizard opens the mouth, while both throat and eyespots might work better at long distance. Meanwhile, throat colouration and eyespots probably work better in different light conditions, which might explain the altitudinal variation in the relative importance of each colour component.

Keywords: colouration; social signals; *Psammodromus algirus*; lizards; altitudinal gradient

Citation: Moreno-Rueda, G.; Reguera, S.; Zamora-Camacho, F.J.; Comas, M. Inter-Individual Differences in Ornamental Colouration in a Mediterranean Lizard in Relation to Altitude, Season, Sex, Age, and Body Traits. *Diversity* **2021**, *13*, 158. https://doi.org/10.3390/d13040158

Academic Editor: Luc Legal

Received: 8 March 2021
Accepted: 5 April 2021
Published: 6 April 2021

Publisher's Note: MDPI stays neutral with regard to jurisdictional claims in published maps and institutional affiliations.

1. Introduction

Colour ornaments are frequent in the animal kingdom, typically involved in social communication [1]. However, understanding the evolution of colour patterns in animals is challenging because it often arises from the interaction of concomitant selective pressures. Sexual selection favours sexual dichromatism and colourful patterns [2], whereas natural selection selects for dull and cryptic colourations [3]. Besides, animals frequently show complex and contrasting colour patterns consisting of multiple colour patches [4]. Multiple ornamental colour patches within individuals may be the result of different selective pressures on each patch [5,6], and so each component of colouration may be related to different individual traits (multiple message hypothesis; e.g., [7–9]). Alternatively, different colour patches may act as redundant signals providing similar information, in this way increasing the reliability of the signal (backup hypothesis, [10,11]).

In reptiles, colouration is the result of a mix of pigments (carotenoids, pterins, and melanin) and structural layers including crystalline platelets [12]. In lizards, the size and spectral characteristics of colour patches have been related to different traits [13]. For example, different characteristics of colour patches can reflect fighting ability [14–16], as well as reproductive status in both females [17–19] and males [20–22], among several other traits.

Understanding the processes that generate intraspecific phenotypic variation is essential in evolutionary ecology. Colouration may vary geographically if selective pressures also vary [23–25]. In this sense, elevational gradients offer a valuable study framework encompassing a considerable environmental variation in a relatively short spatial range [26]. For example, Badyaev [27] described an elevational pattern in birds, in which the strength of sexual selection on colouration decreased with altitude. Badyaev [27] proposed several explanations for this pattern, some of which apply only to species with prolonged parental care. However, it is unknown whether such a pattern is also applicable to reptiles, studies on lizards providing mixed results [28–32].

In the present study, we investigated several sources of inter-individual variation in social colouration of the Mediterranean lizard *Psammodromus algirus* in Sierra Nevada Mountain (Spain). *Psammodromus algirus* is a medium-large lacertid (53–80 mm snout-vent length, SVL, in our study area) that inhabits shrubby habitats in the Mediterranean region of western North Africa, the Iberian Peninsula, and southern France [33]. In the Iberian Peninsula, this species is split into two phylogeographic clades [34], which differ in male colouration during the breeding season (see Figures 1 and 2 in [33]). In the western clade, adult males typically show orange-red colour on the head and throat during the breeding season [19,30,35–37], while young males only exhibit an orange spot in their mouth commissures [38]. These colour patches are absent in females. Meanwhile, social colouration in the eastern clade has been much less studied. In the eastern clade, adult males present an orange spot in their mouth commissures, but not orange colouration in the head and throat [39], just like young males in the western clade. Both adult males and females may exhibit a yellow patch on their throats during the breeding season [39]. Besides, *P. algirus* from both clades display a variable number of blue-ultraviolet eyespots in their flanks, which are more numerous in males than in females, and show little seasonal variation [39,40].

The aim of this study is to examine the sources of inter-individual variation in the colour variables involved in social communication of a southern population within the eastern clade of *P. algirus*. We were interested in the diversity of colour patches in this species and its marked geographical variation. We related several colour traits (throat lightness, chroma, and hue, commissure colour and patch size, and the number of flank eyespots) with:

(1) Morphometric traits (SVL, body mass, and head size). Lizard morphometric traits strongly correlate with fitness. Body size is positively related to reproductive success in males [35,37,41] and females [42]. Head size is typically related to bite force, representing fighting ability, so it is related to social dominance in *P. algirus* [37]. While it is well known that head breeding colouration (orange) indicates body size and fighting ability in the western clade [35,37,41], intervening in communication during agonist encounters, whether that is also the case of colour patterns in the eastern clade remains poorly understood.

(2) Sex. Albeit well reported in the western clade [35,36], sexual dichromatism is understudied in eastern populations (but see [39]). Some degree of dichromatism is expected given that sexual selection is typically stronger in males, especially in polygynous species [2]. While the mating system in the eastern clade is unknown, lizards from the western clade are polygynous; a male's territory overlaps with those of several females [35,37].

(3) Age. In the western clade, adult males are well differentiated from immature males by head colouration, but how colouration patterns vary with the advances of years is still unknown. In fact, in the western clade, it is well-determined that male head colouration varies with body size [35]. However, given that lizards are indeterminate growers, body size increases with age, so it is unclear whether colour signals indicate body size or age in this lizard. Males indicating old age may be preferred by females as their signals would indicate longevity and hence individual quality [43].

(4) Altitude. We examine how social colouration varies with altitude along a 2200-m elevational gradient. In a population of the western clade, lizards from localities separated 650 m in altitude differed in colouration, low-elevation individuals having more saturated colour in throats and more eyespots than high-elevation conspecifics [29,30]. However, whether colouration in populations from the eastern clade similarly varies with altitude remains unknown.

(5) Season. In the western clade, some signals as head and throat patch size and colouration vary seasonally [35], while others as the number of eyespots do not [40]. Seasonal variation in colouration of the eastern clade is poorly known (but see [39]).

Our final goal is to add to the knowledge of the evolution of lizard colouration. Concretely, our main purposes are to understand whether such a variety of signals provide different information on lizard quality or well provide redundant information, as well as to give insights on the sources of geographic variation in social colouration.

Figure 1. The three-dimensional map (bottom panel) is a representation of the Sierra Nevada Mountain and displays the location of the six sites sampled during this study along the elevational gradient: 300 (1), 700 (2), 1200 (3), 1700 (4), 2200 (5), and 2500 (6) m asl. The location of Sierra Nevada in the Iberian Peninsula (**top**, **left**) and an image of the lizard (**top**, **right**) are also shown.

Figure 2. Representation of procedures for quantifying the coloured commissure area, using the software Image J (version 1.60). (**A**) First, we performed lateral photographs of the lizard with the mouth opens; (**B**) Second, we selected the coloured area with the "Colour threshold" tool; (**C**) Third, we measured the area with the "Analyse particles" tool.

2. Material and Methods

2.1. Sampling and General Procedures

Fieldwork was performed in the Sierra Nevada mountain system (SE Spain), where *P. algirus* inhabits from 200 to 2800 m asl, with permission of the Andalusian government and National Park of Sierra Nevada (references GMN/GyB/JMIF and ENSN/JSG/JEGT/MCF). We sampled six localities sited at 300, 700, 1200, 1700, 2200, and 2500 m asl (Figure 1) with a similar structure of vegetation (more details in [44]). Lizards were captured by hand and transported to the lab in cotton bags. Sampling occurred during their activity season in Sierra Nevada, spanning from March to September [44], during the years 2010–2013. We captured 482 adult lizards (males/females per year, 2010: 43 | 58, 2011: 49 | 50, 2012: 65 | 60, 2013: 82 | 75). We measured their SVL with a metal ruler (accuracy 1 mm), body mass with a digital scale (accuracy 0.01 g) and head width with a digital calliper (accuracy 0.01 mm). Sex was determined according to femoral-pore development (more developed in males [29]). Only adults were considered because we were interested in inter- and intra-sexual communication. Given that SVL did not differ between sexes in our study area [45], we considered adults lizards with the minimum body size for which we found gravid females. Notice that body size varies with altitude [45], so this minimum SVL was estimated for each altitude (at 300 m: 53 mm; 700–1700 m: 55 mm; 2200 m: 62 mm; 2500 m: 63 mm). Because lizards were part of a long-term study, they were marked by toe clipping and resampled lizards (~5%) were excluded to avoid pseudoreplication. Toe clipping is frequently used to mark lizards with little impact on welfare and survival [46]. Toes of a subsample (n = 118) were conserved in ethanol and used for age determination using phalanx skeletochronology (detailed methods in [47]). Skeletochronology was a technique widely used to estimate age in reptiles (at the accuracy of a year), which has proven to be very accurate [48].

2.2. Quantification of Colour Patches

Along the four years of the study, we measured throat (gular region) colour with a colourimeter (Minolta CM-2600d). The colourimeter, placed on the lizard's skin, projected three beams of light through a 3-mm-diameter hole. As a result, it took three measures of reflectance and automatically gave the average of each colour component [49]. These components correspond to the L*a*b* colour-space of the Commission Internationale d'Eclairage (CIE 1976 [50]). The device measured the reflectance of the surface on the spectrum of light of the visible range, from 400 to 700 nm wavelengths. We did not study skin reflectance in the ultra-violet (UV) range of the spectrum because the throat has no reflectance peak in the UV range. L*a*b* colour space is a 3-dimensional rectangular colour space. L* axis represents lightness (0 is black, 100 is white); a* axis represents red-green gradient (positive values are red, negative values are green); b* axis represents blue-yellow gradient (positive values are yellow, negative values are blue). From L*a*b* values we determined *chroma* (saturation or purity) as $C^* = [(a^*)^2 + (b^*)^2]^{1/2}$ (measured as the percentage distance from the centre [0] of the colour space to its circumference [100] where pure spectral colours are represented); and hue angle (the "colour" in common parlance) as $H^* = \tan^{-1}(b^*/a^*)$ [51].

In 2011–2013, we additionally recorded the presence or absence, and colour (orange or yellow) if present, of a patch in the mouth commissures, and counted the total number of blue eyespots in the lizard flanks. For those lizards presenting a well-differentiated colour patch in commissures, we took photographs from the right side of the head (consistently with the mouth open; Figure 2). We used a Canon Power Shot SX200 IS digital camera and a graph paper background for size reference. Then, we measured the area of this pigmented patch using the software Image J (version 1.60 [52]). First, we scaled the photos using the graph paper (in mm) and the "Set scale" tool. Then, we adjusted the area of the patch using the "Colour threshold" tool. We afterwards measured the area of each coloured patch with the "Analyse particles" tool (Figure 2). The software measured all the patches delineated and gave the total selected area in mm^2.

2.3. Statistical Analyses

Sample sizes differed among the variables measured (Appendix A). Eyespot number and commissure colour and size were not recorded in 2010. The size of the commissure patch was measured only in lizards in which it was present (50% of lizards). Age was estimated in a random subsample of 118 lizards. For diverse reasons, some data for different variables were lost in several individuals. The fact that not all variables were available in every individual conditioned the statistical analyses we could perform minimising the loss of sample size.

Firstly, we checked for possible outliers in every variable by using Cleveland plots [53]. Secondly, we graphically checked the normality and homoscedasticity of the variables [53]. Throat hue and area of the commissure were log-transformed to match these assumptions. Throat lightness was arcsin-transformed. SVL, body mass and head width were log-transformed to meet homoscedasticity and linearity [54]. Throat chroma and the number of eyespots were not transformed. Although all analyses were carried out with the variables transformed, raw data are shown in graphics and when providing mean values (with SE).

The colour variables measured (number of eyespots, throat lightness, throat chroma, throat hue, presence and colour of commissure patch, and area of the commissure patch) may not be independent among themselves. For this reason, in a first analysis, we examined the bivariate Pearson product-moment correlation between continuous variables, and we tested with Anova whether the value of the colour variables differed among categories of commissure colour (colourless, orange, yellow).

In a second analysis, we carried out a linear model for every colour variable with a continuous distribution, and a multinomial model linked to a logit function for commissure colour. In these models, the predictor variables were SVL, body mass, head width (all log-transformed, continuous), altitude (six levels: 300, 700, 1200, 1700, 2200, 2500 m), season (date when lizards were captured, arranged into three categories: March–May, June–July, August–September), sex (males, females), and year (2010, 2011, 2012, 2013; data for 2010 only available for throat lightness, chroma, and hue). All analyses were carried out with

R 3.6.1 [55]. Multinomial models were performed using the function "multinom" of the package "nnet" v. 7.3–14 [56]. In addition, we applied a model selection approach based on the Akaike Information Criterion (AIC) with the "MuMIn" package [57]. In this analysis, models with a value of ΔAIC less than 2 were chosen [58]. Moreover, given that some colour variables showed some correlation (see below), we repeated the best models including the other variables of colouration as covariates to check the soundness of the findings. Normality and homoscedasticity of every model's residuals were checked [53]. Only significant effects supported by the three statistical approaches (full model, model selection, and best models controlling for other colour variables) were considered as sound. We also checked for interactions between the predictor variables. However, most interactions were non-significant or lacked soundness, so we decided not to show them for the sake of clarity.

Lastly, lacertid lizards typically have indeterminate growth. Consequently, older lizards are also larger. In this way, age and body size may be confounded. To test the effect of age, we repeated the analyses with a subsample of known-age individuals, by including the age estimated with skeletochronology in the models. Given that these models included SVL and age, the effect of the two variables can be disentangled.

3. Results

3.1. Relationship among Colour Parameters

The number of blue eyespots was unrelated to the other colouration variables (throat chroma, lightness and hue, and area of the commissure), but lizards with orange commissures had significantly more eyespots ($F_{2,271} = 31.54$, $p < 0.001$; Table 1). This finding seems to be linked to the fact that males have more eyespots and a higher probability to present orange commissures (see below). Indeed, when sex was introduced as a predictor, the relationship between number of eyespots and colour of commissures disappeared ($F_{2,270} = 1.49$, $p = 0.23$; effect of sex: $F_{1,270} = 39.15$, $p < 0.001$). Meanwhile, throat colouration parameters were interrelated; chroma and lightness were negatively correlated, while hue was positively correlated with lightness and so negatively related to chroma (Table 1). Nonetheless, correlation coefficients were relatively low ($|r| \leq 0.51$). The size of commissure patch was positively correlated with throat chroma, and negatively with hue and lightness (Table 1). In addition, individuals with a commissure patch (orange or yellow) had more saturated throats ($F_{2,274} = 23.31$, $p < 0.001$; Table 1). The reverse occurred for throat lightness ($F_{2,274} = 17.96$, $p < 0.001$; Table 1). Throat hue was unrelated to commissure colour ($F_{2,274} = 0.24$, $p = 0.79$). Lizards with orange commissures tended to have larger commissure patches than lizards with yellow commissures (0.74 ± 0.07 vs. 0.48 ± 0.16 mm^2; $F_{1,111} = 3.14$, $p = 0.079$). Therefore, mouth commissure and throat colourations were seemingly interrelated.

3.2. Correlates of Colouration

The full model (Table 2) showed that the number of eyespots was maximal at a low elevation and minimal at mid-elevation (Figure 3a). Males had more eyespots than females (Figure 3b). Moreover, lizards with larger heads showed more eyespots (Figure 3c). The number of eyespots did not significantly vary with season or body size (SVL or mass). According to a model selection approach, the best model was that including altitude, sex, and head size (Table 3). The second-best model included those variables plus SVL (ΔAIC = 1.79), which, however, failed to significantly explain variation in the number of eyespots.

Throat lightness significantly varied with altitude (Table 2). Concretely, lizards from localities at 2200 and 2500 m had darker throat than those from lower altitudes (Figure 4a). Lizard throat was darker at the end of the breeding season (Figure 4b). Lizards with larger SVL and heads showed darker throats than smaller lizards (Table 2; Figure 4c; data not shown for SVL). Besides, throat lightness varied among years (Table 2). The four best models (ΔAIC < 2) included altitude, season, year, and head size as predictors (Table 3). Given that throat lightness was correlated with throat chroma and hue, we repeated the best model including these two variables as predictors to control for them. The resulting model was qualitatively very similar to the best model (data not shown), with the exception that SVL was no longer significantly related to throat lightness. SVL was not included in the fourth best model (Table 3), so there was little support for this variable being related to throat lightness.

Throat chroma increased with altitude (Figure 5a) and decreased with the advance of the season (Figure 5b; Table 2). Throat chroma increased with SVL and head size (Figure 5c, data not shown for head size). Besides, throat chroma showed interannual variation (Table 2). With the model selection, three models had ΔAIC < 2, all including altitude, season, year, SVL, and head size as significant predictors of throat chroma (Table 3). When controlling for throat lightness and hue, results were similar, but head size was no longer significant, and a marginally significant effect of body mass emerged (data not shown).

Table 1. Correlations among the continuous variables (provided sample size -in the subscript-, correlation coefficient and p-value are provided), and average values ($\pm$SE) for each category of commissure colour (no colour, yellow, or orange). Sample sizes for each category of commissure colour between brackets (for eyespots of lizards with orange commissures, the sample size was 103). Raw data are shown, but statistical tests were carried out with transformed variables when necessary. In bold significant relationships or differences. Different superscripts indicate significant differences according to an unequal N HSD post hoc test.

	Correlations				Commissure Colour		
	Throat Lightness	**Throat Chroma**	**Throat Hue**	**Commissure Area**	**No (138)**	**Yellow (33)**	**Orange (106)**
Eyespots	$r_{362} = 0.02$ $p = 0.71$	$r_{362} = 0.03$ $p = 0.51$	$r_{362} = 0.08$ $p = 0.12$	$r_{110} = 0.04$ $p = 0.71$	**3.64 ^{a} $\pm$ 0.18**	**3.85 ^{a} $\pm$ 0.44**	**5.79 ^{b} $\pm$ 0.20**
Throat Lightness		**$r_{473} = -0.51$ $p < 0.001$**	**$r_{473} = 0.21$ $p < 0.001$**	**$r_{114} = -0.30$ $p < 0.001$**	**81.28 ^{a} $\pm$ 0.37**	**76.09 ^{b} $\pm$ 1.12**	**79.57 ^{c} $\pm$ 0.44**
Throat Chroma			**$r_{473} = -0.13$ $p = 0.006$**	**$r_{114} = 0.46$ $p < 0.001$**	**12.20 ^{a} $\pm$ 0.36**	**20.10 ^{b} $\pm$ 1.58**	**17.19 ^{b} $\pm$ 0.92**
Throat Hue				$r_{114} = -0.18$ $p = 0.06$	88.86 $\pm$ 1.09	88.87 $\pm$ 2.10	87.70 $\pm$ 1.08

Table 2. Full models for each colour variable. In bold, predictors with a significant effect. Degree of freedom (df) as well as F-value for lineal models and χ^2 for multinomial model are shown. * $p < 0.05$, ** $p < 0.01$, *** $p < 0.001$.

			Throat				Commissure		
	Eyespots			Lightness	Chroma	Hue	Colour		Area	
	Df	*F*-Value	df	*F*-Value	*F*-Value	*F*-Value	df	χ^2	df	*F*-Value
Altitude	5, 333	**3.88 ****	5, 430	**8.48 *****	**3.25 ****	**6.12 *****	10	4.08	5, 97	0.75
Season	2, 333	0.08	2, 430	**12.79 *****	**10.93 *****	**8.16 *****	4	**10.38 ***	2, 97	**4.96 ****
Year	2, 333	1.18	3, 430	**6.48 *****	**5.93 *****	**2.98 ***	4	1.92	2, 97	**6.58 ****
Sex	1, 333	**57.62 *****	1, 430	2.98	0.73	1.77	2	**90.72 *****	1, 97	2.23
Mass	1, 333	0.71	1, 430	0.99	2.83	0.31	2	0.99	1, 97	2.67
SVL	1, 333	1.44	1, 430	**4.58 ***	**8.50 ****	1.12	2	0.73	1, 97	1.66
Head	1, 333	**6.78 ***	1, 430	**8.94 ****	**4.86 ***	1.12	2	**8.99 ***	1, 97	0.15

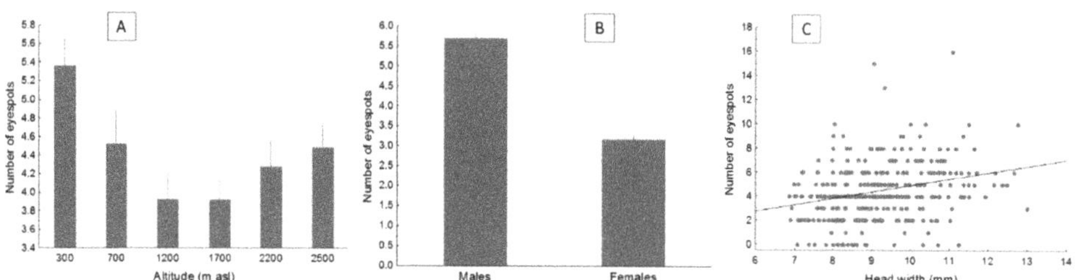

Figure 3. The average number of eyespots (with SE) in the flanks of the lizard *P. algirus*, according to altitude (**A**) and sex (**B**), and the relationship between the number of eyespots and head width (**C**). Notice that raw data are shown, but statistical analyses were performed with transformed data when necessary.

Table 3. Models chosen by model selection approach based on AIC. In bold, predictors with a significant effect.

Dependent Variable	Predictors	df	AICc	ΔAIC	Weight
Eyespots	**Altitude + Sex + Head**	9	1437.01	0	0.29
	Altitude + Sex + Head + SVL	10	1438.80	1.79	0.12
Throat Lightness	**Altitude + Season + Year + Head + SVL** + Sex	15	−1161.73	0	0.31
	Altitude + Season + Year + Head + SVL + Sex + Mass	16	−1160.61	1.13	0.18
	Altitude + Season + Year + Head + SVL	14	−1160.28	1.45	0.15
	Altitude + Season + Year + Head + Sex	14	–1159.82	1.91	0.12
Throat Chroma	**Altitude + Season + Year + Head + SVL** + Mass	15	2863.91	0	0.31
	Altitude + Season + Year + Head + SVL	14	2864.35	0.45	0.25
	Altitude + Season + Year + Head + SVL + Mass + Sex	16	2865.30	1.40	0.15
Throat Hue	**Altitude + Season + Year + Sex**	13	–648.44	0	0.25
	Altitude + Season + Year + Sex + Head	14	–646.98	1.46	0.12
	Altitude + Season + Year + Sex + Mass	14	–646.49	1.95	0.10
Commissure Colour	**Season + Sex + Head** + Mass	12	272.67	0	0.41
	Season + Sex + Head + SVL	12	273.09	0.42	0.33
Commissure Area	**Year + Season + Mass** + Head	8	198.57	0	0.15
	Year + Season + Sex + Mass	8	198.74	0.17	0.14
	Year + Season + Sex + SVL + Mass	9	198.87	0.30	0.13
	Year + Season + SVL + Head	8	199.25	0.68	0.11
	Year + Season + Sex + SVL	8	199.75	1.18	0.08
	Year + Season + Sex + **Mass** + Head	9	199.81	1.24	0.08
	Year + Season + SVL + Mass + Head	9	200.13	1.55	0.07
	Year + Season + Sex + **SVL** + Head	9	200.22	1.65	0.06

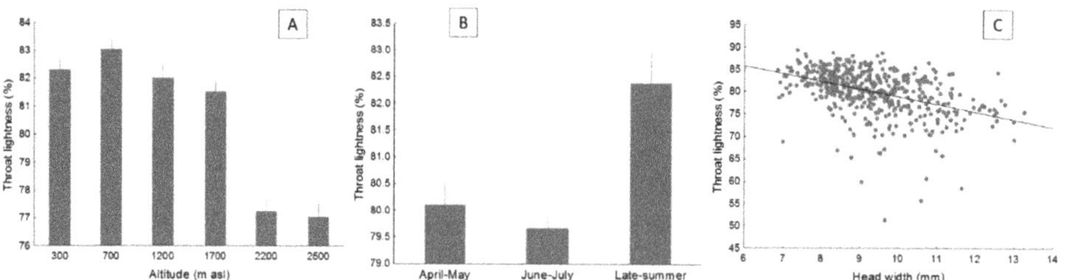

Figure 4. Average values of throat lightness (with SE) in the lizard *P. algirus*, according to altitude (**A**) and season (**B**), and the relationship between throat lightness and head width (**C**). Notice that raw data are shown, but statistical analyses were performed with transformed data when necessary.

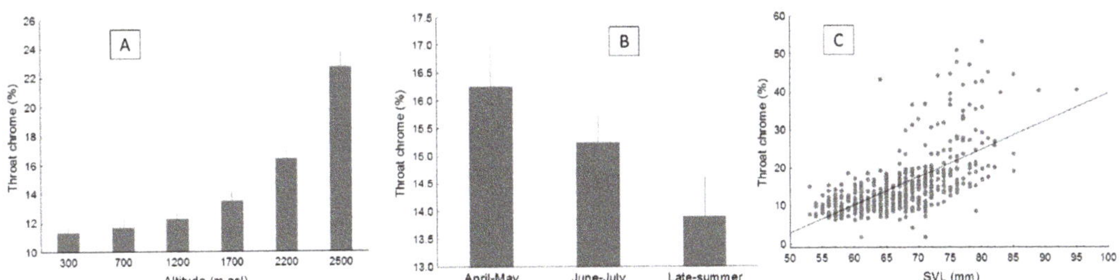

Figure 5. Average values of throat chroma (with SE) in the lizard *P. algirus*, according to altitude (**A**) and season (**B**), and the relationship between throat chroma and SVL (**C**). Notice that raw data are shown, but statistical analyses were performed with transformed data when necessary.

According to the full model, throat hue showed covariation with altitude, season, and year, but not with sex or lizard morphology (Table 2). Throat hue tended to decrease with altitude and increased with the advance of the season (data not shown for simplicity). The model selection provided similar results, but, the best models also included sex, which significantly explained part of the variation in throat hue (Table 3). The inclusion of throat lightness and chroma in the model did not alter significantly the results (not shown).

The colour of the commissure differed between sexes, males usually having orange commissures, while females had yellow or no coloured commissure (Figure 6a; Table 2). The frequency of lizards with orange commissures decreased with the advance of the breeding season (Figure 6b; Table 2). Lastly, individuals with coloured commissure (either orange or yellow) had larger heads than individuals without commissure (Figure 6c, Table 2). No other variable was significantly related to commissure colour (Table 2). Model selection approach selected for two models including sex, season, and head size as significant predictors (Table 3).

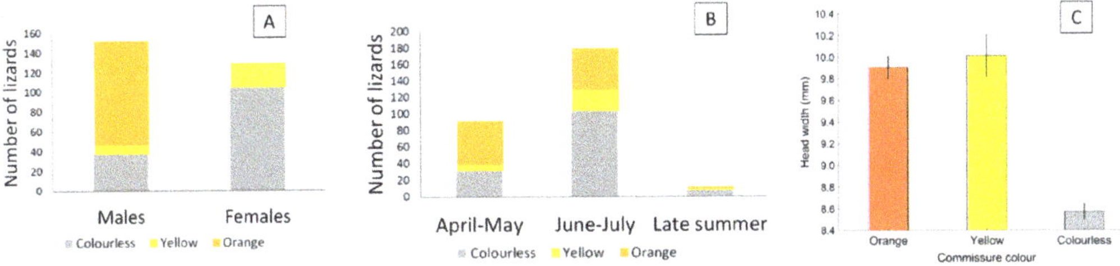

Figure 6. Frequency of *P. algirus* lizards with colourless, yellow or orange commissure according to sex (**A**) and season (**B**), and head width (with SE) of lizards according to commissure colour (**C**).

The area of the coloured commissure varied significantly with season and year (Table 2), tending to decrease with the advance of the breeding season (Figure 7). Eight models were selected by model selection. All models included season and year as significant predictors. No model included altitude. The remaining variables were included in five models, being significant in some, but not in others (Table 3). A model including throat lightness, chroma, and hue did not alter significantly these results (not shown).

3.3. The Effect of Age

As expected, SVL increased with age ($F_{4,\,113} = 14.06$, $p < 0.001$; Appendix B). We found that older individuals were more likely to have coloured commissures ($\chi^2_6 = 12.83$, $p = 0.046$; Figure 8a). Moreover, throat chroma increased with age (Figure 8b). This last result was confirmed by model selection. Two models were selected including the variables age + body mass (AICc = 678.33) and age + body mass + head size (AICc = 679.14). Notice that the two models included age, but not SVL, as a predictor of throat chroma. Previous analyses showed that throat chroma was the only colour variable that increased with SVL (Table 4). However, the present findings suggest that throat chroma is influenced by age rather than by body size.

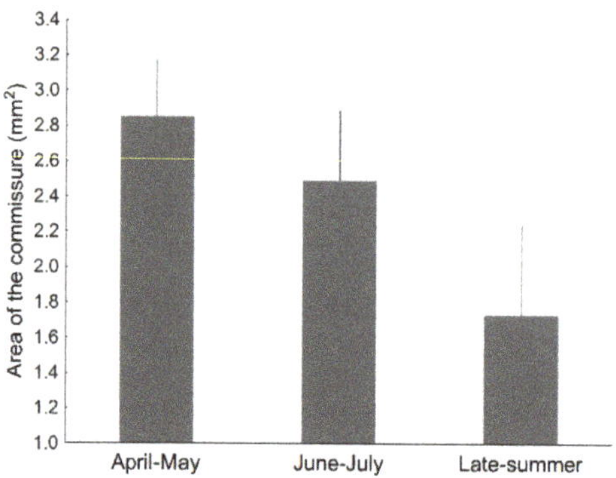

Figure 7. Average values of commissure area (with SE) in the lizard *P. algirus*, according to season.

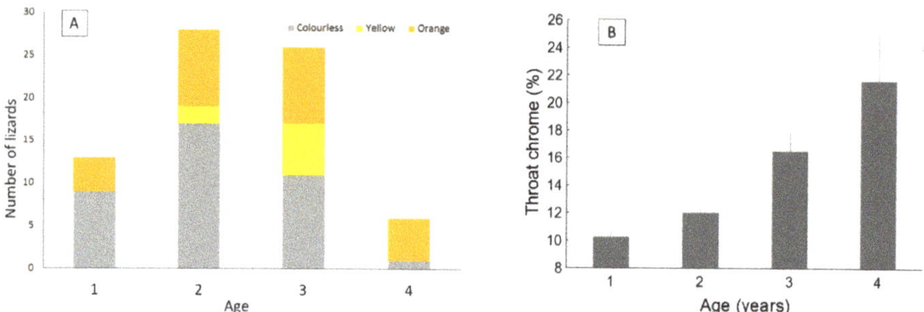

Figure 8. Frequency of lizards with colourless, yellow or orange commissure according to age (**A**) and average (with SE) throat chroma according to age (**B**).

Table 4. Summary of the main results. In bold are the clear results, whereas somewhat dubious results are not in bold.

	Altitude	Season	Year	Sex	SVL	Mass	Head
Eyespots	**U-shaped**	No	No	**More in males**	No	No	**Increase**
Throat Lightness	**Darker at the highest altitudes**	**Lighter at the end of the season**	**Yes**	No	Decrease	No	**Decrease**
Throat Chroma	**Linear increase**	**Linear decrease**	**Yes**	No	**Increase (with age)**	No	Increase
Throat Hue	**Decrease**	**Linear increase**	Yes	Yes	No	No	No
Commissure Colour	No	**Less orange lizards with advanced season**	No	**Orange, males; yellow or colourless, females**	No	No	**Larger in coloured lizards**
Commissure Area	No	**Linear decrease**	**Yes**	Larger in males	No	No	No

4. Discussion

Table 4 summarises the conclusions based on the main results. The findings in this study suggest that blue eyespots function independently from colour patches in the commissure and throat, while throat and commissure colouration were interrelated. Moreover, eyespots and both commissure and throat varied in different ways with altitude and season. Eyespots varied with altitude following a U-shaped pattern, while throat colour increased in saturation (chroma) with altitude. Also, throat colouration was darker in the highest elevations, where lizards are typically darker [59]. While eyespots were permanent colour patches (as in the western clade [40]), commissure and throat colouration varied with the season. At the beginning of the breeding season, we found more lizards with coloured commissures and throat colour was darker and more saturated. Similar findings were reported for a northern population of the same phylogenetic clade [39], which suggests that throat and commissure patches are involved in communication during breeding. Similarly, in the western clade, orange head colouration in adult males and coloured commissures in young males are present during the breeding season [19,35,36,38].

We found some degree of sexual dichromatism; males typically had more eyespots and orange commissures (which were yellow or colourless in females). Hence, although sexual dichromatism was not as pronounced as in the western clade, a slight dichromatism is still present in the eastern clade. Throat saturation and the presence of coloured commissures was indicative of age. Meanwhile, the number of eyespots, presence of a coloured commissure, and throat saturation, all were indicators of head size, and thus of fighting ability. Therefore, although colour patches changed with altitude and season in different ways (probably reflecting different subjacent colour production costs and mechanisms and/or selective pressures), they seem, in general, to be redundant indicating the same traits of lizards: sex, fighting ability, and age. One possibility is that these signals are used in different contexts, indicating the same traits at different distances or in different light conditions.

4.1. Colouration as Indicators of Fighting Ability, Sex, and Age

All colour patches measured (eyespots, commissure and throat) seem indicators of head size (Table 4), which is known as a good indicator of fighting ability [60]. Lizards with greater fighting ability are more successful in defending their territories and hence obtain higher mating success [37]. Several studies suggest that blue-ultraviolet colouration, as that present in eyespots, is related to fighting ability [16,61,62], but pigmentary colourations may also serve as indicators of fighting ability [63]. During a contest, signals correlated with fighting ability may help individuals to assess the relative competitive ability of rivals and so to avoid being involved in a costly physical combat [2]. To be informative, such signals should be honest indicators of fighting ability. Colouration may be a reliable indicator of fighting ability when maintained by social interactions [64]. Indeed, in the western clade of *P. algirus*, orange head colouration is a good indicator of social dominance, but orange-headed lizards are also more often involved in fighting, so subordinate lizards would pay a cost if vividly coloured [41]. Moreover, throat and commissure colouration might act as amplifiers of head size. This may be especially important in commissure colouration, which is displayed only when the mouth is open, hence allegedly showing the intention of biting and amplifying the perceived mouth size [65]. Furthermore, in the western clade, several costs have been associated with male nuptial colouration, such as reduced immune capacity, increased risk of ectoparasitism and ultimately reduced survival, which could serve to maintain honesty [30,35,36].

While lizards from western populations are strongly dichromatic, sexual dichromatism in our study population was reduced to commissure colour (orange in males, yellow or colourless in females) and the number of eyespots (more numerous in males). Therefore, commissure colour seems to intervene in sex recognition. The fact that females display yellow throat and blue eyespots similarly to males suggests that females also use colour patches in social communication. Female lizards frequently display colour patches, usually related to receptiveness to breed [17–19]. However, alternative explanations, typically poorly explored, are possible. Females could use throat colouration (and yellow commissures) to indicate fighting ability to rivals. Females could also indicate some type of individual quality to potential mates [66]. However, colouration in females might simply be the result of correlated selection in males [67]. Anyway, the reduced sexual dichromatism in the eastern clade suggests that sexual selection on male traits is weaker in the eastern than in the western clade.

In Sierra Nevada, older individuals expressed both the yellow patch on the throat and the orange commissures more frequently than young lizards. Given that this is a trans-sectional study, we cannot disentangle whether colourful lizards lived longer (duller lizards selectively disappearing from the population as they aged), or lizards invested more in colouration as they aged. In either case, these colour patterns would indicate higher survival capability of the bearer. Younger lizards may be selected for concealment, mimicking females, hence not developing the yellow patch or the orange commissure until they grow large enough to compete with older males [38,41]. However, dominant males may detect female-mimicking males using chemosensorial cues [68]. Alternatively, colour cues indicating lizard age could be used to evaluate the survival prospects of potential mates [43]. In order for signals of survival ability to be honest, they should be costly to produce or maintain for their bearers [69]. For instance, in the western clade, conspicuous orange head colourations make individuals more visible to predators [70], so only high-quality lizards may survive older.

4.2. Altitudinal Variation in Colouration

We report that eyespots and throat colouration followed different altitudinal patterns; the number of blue eyespots was highest at low elevation and then followed a U-shaped trend with altitude, while saturation of throat colouration linearly increased with elevation. These discrepant patterns are hard to explain on the only basis of altitudinal variation in sexual selection pressure. Moreover, altitudinal colour variation was similar in males and females, supporting the idea that it is provoked by natural selection, not by sexual selection [71]. Our findings contrast with those reported for a population of the western clade in central Spain, where *P. algirus* lizards have less saturated throats at higher altitude [29,30]. In our study population, lizard throat was darker at high elevations, which has also been reported in other lizard species [28,32]. Darker colouration with ascending elevation is a likely consequence of high dermal melanin [59].

This pattern of altitudinal variation in lizard colour signals and, especially, why eyespots and throat colouration covaried differently with elevation, requires an explanation in which both colour signals are differentially affected by selective pressures. Blue eyespots are structural colourations, mainly produced by the combination of a thick and well-arranged layer of iridophores and basal eumelanin. Meanwhile, yellow and orange colourations are produced by pigments such as carotenoids and pterins. We discuss several hypotheses that could explain the altitudinal patterns reported.

(1) Population density might affect the investment in social communication. In more densely populated zones, social encounters should be more frequent, conducing to increased contests. Given that colour signals in *P. algirus* apparently inform about fighting ability, one could expect more investment in social signals in zones with denser populations. However, this hypothesis is not supported, as density was maximal at mid-elevation, and minimal at lowlands [44], where the number of eyespots was the highest. Therefore, altitudinal variation in population density did not covary with either variation in the number of eyespots or throat saturation.

(2) We could also expect a trade-off between investment in signals and in self-maintenance [2]. Therefore, in zones where lizards invest more in longevity, social signals are expected to be less expressed (i.e., the pattern of longevity and signal intensity should be inverse). This hypothesis was not supported either, as longevity followed a U-shape with altitude [72], hence showing a pattern not consistent with a trade-off between self-maintenance and investment in colouration. Eyespots also followed a U-shaped pattern and throat colouration increased with elevation.

(3) Temperature is the main environmental factor that varies with elevation. Colder temperatures at higher elevations may limit activity, especially for ectotherms [44,73]. Moreover, elevated temperatures may favour sexual selection [74]. Although this

could partially explain the highest number of eyespots at low elevations, hardly could it explain why investment in throat colouration increases with ascending altitude.

(4) Several colour traits in lizards are sensitive to parasites [36,75–80]. However, parasites follow a complex pattern with altitude in our study system, mite abundance decreasing, while haemoparasites prevalence increases, with ascending altitude [81]. Still, different types of colouration may be related to different types of parasites [82].

(5) Food availability increases with elevation in our study population [83]. Chromatic properties of lizard skin correlate with pigment density, highly saturated colour patches having more pigment density than paler and duller ones [84]. Therefore, if throat pigment concentration depends on food (e.g., if they are carotenoid-dependent), the increase in food availability could explain the highest investment in throat colouration with altitude [85,86]. The altitudinal pattern for eyespots would be different as structural colour are presumed to be less affected by food availability.

(6) Lizards at higher elevations suffer less oxidative damage than low-elevation ones [87,88]. Although we do not know if the yellow colouration in the throat is mediated by carotenoids or by pterins, both pigments may have antioxidant properties mainly mediated throughout a regulatory effect on the immune system [89]. Therefore, lizards from a higher elevation, exposed to a less oxidant ambient, could invest more in pigment-mediated social signals [90]. Meanwhile, the structural colouration of blue eyespots could be unaffected by oxidative stress.

(7) Altitudinal variation in female preferences for different colour traits in males could explain altitudinal variation in colouration [91] (also see [92,93]). However, this would hardly explain the low sexual dichromatism along the altitudinal gradient [71].

(8) The efficiency of colour signals depends on the context where the visual stimulus is produced, as environmental conditions also affect the dispersion of the signal [94]. Blue-ultraviolet colours (short wavelengths) are more effective in partially covered habitats, while yellow-orange colourations are more effective in open areas [94]. Low elevation sites, where eyespots are more numerous, are composed of Mediterranean forests, with low arboreal cover, loose bushes, and a matrix of open and forestry zones. In this type of habitat, short-length colours would be favoured. Meanwhile, high elevation sites are above the treeline, with a habitat composed of short and compact scrubs. In these sites, by contrast, yellow colouration might be favoured.

4.3. Comparison with the Western Clade

In the western clade, testosterone produces orange heads in adult males, but only an orange commissure in young males [38]. To produce or have an orange head is costly, as coloured individuals are more implied in fights, and colouration increases parasite susceptibility and reduces survival [30,35,36,41]. Why do males in the eastern clade not have orange heads during the breeding season? One possibility is paedomorphism, if adult lizards are retaining their youthful characteristics. This may occur because environmental conditions in the zones inhabited by the eastern clade favour an augment in the costs and/or decrease in the benefits associated with orange colouration. In the western clade, highly coloured males have larger home ranges that overlap with more females and so are more successful in mate acquisition [35,37,41]. The eastern clade inhabits more arid zones, with less plant cover and probably less food availability [42,95]. This might influence the social behaviour of this lizard (which is unstudied in the eastern clade), needing larger territories in which the control of several female territories is more difficult, and the contact with other males is rarer. Therefore, both intra- and intersexual selection could be lower in the eastern clade and so the benefits of an orange head could decrease. It is also possible that lizards in the eastern clade are more exposed to predators, given that they live in an opener environment [95]. Higher risk of predation would select against colourful males [96]. These questions remain open and show that *P. algirus* lizards offer a valuable opportunity to study and understand the evolution of nuptial colouration in lizards.

Author Contributions: Conceptualization, G.M.-R.; methodology, G.M.-R., S.R. and M.C.; resources: G.M.-R.; investigation: S.R., F.J.Z.-C. and M.C.; data curation: S.R., M.C. and G.M.-R.; formal analysis, G.M.-R. and S.R.; writing, original draft, G.M.-R. and S.R.; writing, additional inputs, F.J.Z.-C. and M.C.; supervision, G.M.-R.; project administration, G.M.-R.; funding acquisition, G.M.-R. All authors have read and agreed to the published version of the manuscript.

Funding: This work was funded by the Spanish government (Ministerio de Ciencia e Innovación) with funds from the European Union (project CGL2009-13185). FJZC (AP2009-3505) and SR (AP2009-1325) were supported by two pre-doctoral grants from the Ministerio de Ciencia e Innovación (FPU programme).

Institutional Review Board Statement: The research was conducted under both Junta de Andalucía and National Park of Sierra Nevada research permits (references GMN/GyB/JMIF, ENSN/JSG/JEGT/MCF, ENSN/JSG/BRL/MCF, SGMN/GyB/JMIF, and SSA/SI/MD/ps) issued to the authors.

Informed Consent Statement: Not applicable.

Data Availability Statement: The data presented in this study are openly available in FigShare at https://doi.org/10.6084/m9.figshare.14376698.v1, accessed on 6 April 2021.

Acknowledgments: We thank the personnel from the Espacio Natural de Sierra Nevada for their constant support.

Conflicts of Interest: The authors declare no conflict of interest.

Appendix A

Table A1. Sample sizes per sex (males | females) and per altitude for each variable.

	300	700	1200	1700	2200	2500	Total
SVL	61 | 45	20 | 36	19 | 30	37 | 36	42 | 50	60 | 46	482
Body Mass	59 | 44	20 | 34	18 | 30	36 | 36	42 | 50	60 | 46	475
Head Width	60 | 43	19 | 29	16 | 29	37 | 33	39 | 48	57 | 45	460
Age	11 | 12	8 | 11	10 | 10	10 | 9	9 | 7	11 | 10	118
Eyespots	44 | 28	19 | 29	16 | 23	37 | 32	32 | 37	43 | 31	371
Throat Colour (LCH)	59 | 45	19 | 35	17 | 30	36 | 35	42 | 49	60 | 46	473
Commissure Patch Colour	33 | 22	11 | 15	14 | 19	32 | 25	27 | 24	35 | 24	281
Commissure Patch Size	16 | 1	5 | 2	5 | 2	20 | 6	18 | 5	25 | 9	114

Appendix B

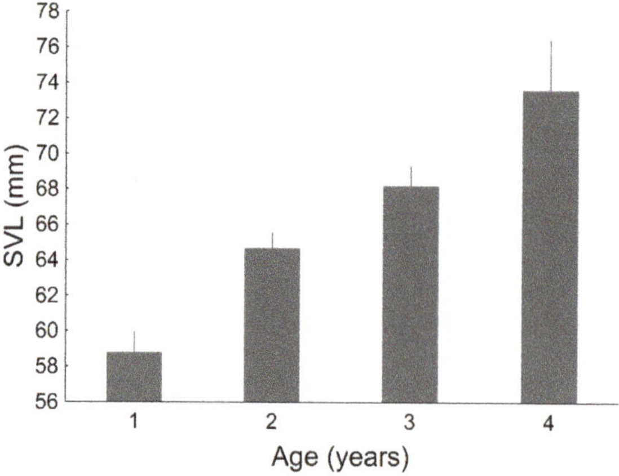

Figure A1. Average snout-vent length (SVL) of *P. algirus* lizards according to age estimated with skeletochronology. Lines indicate the starndard error.

References

1. Bradbury, J.W.; Vehrencamp, S.L. *Principles of Animal Communication*; Sinauer: Sunderland, MA, USA, 2011.
2. Andersson, M. *Sexual Selection*; Princeton University Press: Princeton, NJ, USA, 1994.
3. Stevens, M.; Merilaita, S. (Eds.) *Animal Camouflage: Mechanisms and Function*; Cambridge University Press: Cambridge, UK, 2011.
4. Pérez i de Lanuza, G.; Font, E. The evolution of colour pattern complexity: Selection for conspicuousness favours contrasting within-body colour combinations in lizards. *J. Evol. Biol.* **2016**, *29*, 942–951. [CrossRef]
5. Andersson, S.; Pryke, S.R.; Örnborg, J.; Lawes, M.J.; Andersson, M. Multiple receivers, multiple ornaments, and a trade-off between agonistic and epigamic signaling in a widowbird. *Am. Nat.* **2002**, *160*, 683–691. [CrossRef]
6. Candolin, U. The use of multiple cues in mate choice. *Biol. Rev.* **2003**, *78*, 575–595. [CrossRef] [PubMed]
7. Martín, J.; López, P. Multiple color signals may reveal multiple messages in male Schreiber's green lizards, *Lacerta schreiberi*. *Behav. Ecol. Sociobiol.* **2009**, *63*, 1743–1755. [CrossRef]
8. Plasman, M.; Reynoso, V.H.; Nicolás, L.; Torres, R. Multiple colour traits signal performance and immune response in the Dickerson's collared lizard *Crotaphytus dickersonae*. *Behav. Ecol. Sociobiol.* **2015**, *69*, 765–775. [CrossRef]
9. Zamora-Camacho, F.J.; Comas, M. Beyond sexual dimorphism and habitat boundaries: Coloration correlates with morphology, age, and locomotor performance in a toad. *Evol. Biol.* **2019**, *46*, 60–70. [CrossRef]
10. Møller, A.P.; Pomiankowski, A. Why birds got multiple ornaments? *Behav. Ecol. Sociobiol.* **1993**, *32*, 167–176. [CrossRef]
11. Johnstone, R.A. Multiple displays in animal communication: "Backup signals" and "multiple messages". *Philos. Trans. R. Soc. B* **1996**, *351*, 329–338.
12. Grether, G.F.; Kolluru, G.R.; Nersissian, K. Individual colour patches as multicomponent signals. *Biol. Rev.* **2004**, *79*, 583–610. [CrossRef]
13. Olsson, M.; Stuart-Fox, D.; Ballen, C. Genetics and evolution of colour patterns in reptiles. *Semin. Cell Dev. Biol.* **2013**, *24*, 529–541. [CrossRef] [PubMed]
14. Olsson, M. Nuptial colouration in the sand lizard, *Lacerta agilis*: An intra-sexually selected cue to fighting ability. *Anim. Behav.* **1994**, *48*, 607–613. [CrossRef]
15. Whiting, M.J.; Stuart-Fox, D.M.; O'Connor, D.; Firth, D.; Bennett, N.C.; Blomberg, S.P. Ultraviolet signals ultra-aggression in a lizard. *Anim. Behav.* **2006**, *72*, 353–363. [CrossRef]
16. Bajer, K.; Molnár, O.; Török, J.; Herczeg, G. Ultraviolet nuptial colour determines fight success in male European green lizards (*Lacerta viridis*). *Biol. Lett.* **2011**, *7*, 866–868. [CrossRef] [PubMed]
17. Baird, T.A. Reproductive coloration in female collared lizards, *Crotophytus collaris*, stimulates courtship by males. *Herpetologica* **2004**, *60*, 337–348. [CrossRef]
18. Salica, M.J.; Halloy, M. Nuptial coloration in female *Liolaemus quilmes* (Iguania: Liolaemidae): Relation to reproductive state. *Rev. Esp. Herpetol.* **2009**, *23*, 141–149.
19. Stuart-Fox, D.; Goode, J.L. Female ornamentation influences male courtship investment in a lizard. *Front. Ecol. Evol.* **2014**, *2*, 2. [CrossRef]
20. Díaz, J.A.; Alonso-Gomez, A.L.; Delgado, M.J. Seasonal variation of gonadal development, sexual steroids, and lipid reserves in a population of the lizard *Psammodromus algirus*. *J. Herpetol.* **1994**, *28*, 199–205. [CrossRef]
21. Hews, D.K.; Moore, M.C. Influence of androgens on differentiation of secondary sex characters in tree lizards, *Urosaurus ornatus*. *Gen. Comp. Endocrinol.* **1995**, *97*, 86–102. [CrossRef]
22. Cox, R.M.; Zilberman, V.; John-Alder, H.B. Testosterone stimulates the expression of a social color signal in Yarrow's Spiny Lizard, *Sceloporus jarrovii*. *J. Exp. Zool. A* **2008**, *309*, 505–514. [CrossRef]
23. Macedonia, J.M. Habitat light, colour variation, and ultraviolet reflectance in the Grand Cayman anole, *Anolis conspersus*. *Biol. J. Linn. Soc.* **2001**, *73*, 299–320. [CrossRef]
24. Pérez i de Lanuza, G.; Carretero, M.A.; Font, E. Intensity of male-male competition predicts morph diversity in a color polymorphic lizard. *Evolution* **2017**, *71*, 1832–1840. [CrossRef]
25. Moreno-Rueda, G.; González-Granda, L.G.; Reguera, S.; Zamora-Camacho, F.J.; Melero, E. Crypsis decreases with elevation in a lizard. *Diversity* **2019**, *11*, 236. [CrossRef]
26. Keller, I.; Alexander, J.M.; Holderegger, R.; Edwards, P.J. Widespread phenotypic and genetic divergence along altitudinal gradients in animals. *J. Evol. Biol.* **2013**, *26*, 2527–2543. [CrossRef] [PubMed]
27. Badyaev, A.V. Altitudinal variation in sexual dimorphism: A new pattern and alternative hypotheses. *Behav. Ecol.* **1997**, *8*, 675–690. [CrossRef]
28. Leaché, A.D.; Helmer, D.-S.; Moritz, C. Phenotypic evolution in high-elevation populations of western fencce lizards (*Sceloporus occidentalis*) in the Sierra Nevada Mountains. *Biol. J. Linn. Soc.* **2010**, *100*, 630–641. [CrossRef]
29. Iraeta, P.; Monasterio, C.; Salvador, A.; Díaz, J.A. Sexual dimorphism and interpopulation differences in lizard hind limb length: Locomotor performance or chemical signalling? *Biol. J. Linn. Soc.* **2011**, *104*, 318–329. [CrossRef]
30. Llanos-Garrido, A.; Díaz, J.A.; Pérez-Rodríguez, A.; Arriero, E. Variation in male ornaments in two lizard populations with contrasting parasite loads. *J. Zool.* **2017**, *303*, 218–225. [CrossRef]
31. Martín, J.; Zamora-Camacho, F.J.; Reguera, S.; López, P.; Moreno-Rueda, G. Variations in chemical sexual signals of *Psammodromus algirus* lizards along an elevation gradient may reflect altitudinal variation in microclimatic conditions. *Sci. Nat.* **2017**, *104*, 16. [CrossRef]

32. Ortega, J.; Martín, J.; Crochet, P.-A.; López, P.; Clobert, J. Seasonal and interpopulational phenotypic variation in morphology and sexual signals of *Podarcis liolepis* lizards. *PLoS ONE* **2019**, *14*, e0211686. [CrossRef]
33. Salvador, A. Lagartija colilarga—*Psammodromus algirus* (Linnaeus, 1758). In *Enciclopedia Virtual de los Vertebrados Españoles*; Salvador, A., Marco, A., Eds.; Museo Nacional de Ciencias Naturales: Madrid, Spain, 2015.
34. Carranza, S.; Harris, D.J.; Arnold, E.N.; Batista, V.; Gonzalez De La Vega, J.P. Phylogeography of the lacertid lizard, *Psammodromus algirus*, in Iberia and across the Strait of Gibraltar. *J. Biogeogr.* **2006**, *33*, 1279–1288. [CrossRef]
35. Díaz, J.A. Breeding coloration, mating opportunities, activity, and survival in the lacertid lizard *Psammodromus algirus*. *Can. J. Zool.* **1993**, *71*, 1104–1110. [CrossRef]
36. Salvador, A.; Veiga, J.P.; Martín, J.; López, P.; Abelenda, M.; Puerta, M. The cost of producing a sexual signal: Testosterone increases the susceptibitily of male lizards to ectoparasitic infestation. *Behav. Ecol.* **1996**, *7*, 145–150. [CrossRef]
37. Salvador, A.; Veiga, J.P. Male traits and pairing success in the lizard *Psammodromus algirus*. *Herpetologica* **2001**, *57*, 77–86.
38. Salvador, A.; Veiga, J.P.; Martín, J.; López, P. Testosterone supplementation in subordinate, small male lizards: Consequences for aggressiveness, color development, and parasite load. *Behav. Ecol.* **1997**, *8*, 135–139. [CrossRef]
39. Carretero, M.A. Sources of colour pattern variation in Mediterranean *Psammodromus algirus*. *Neth. J. Zool.* **2002**, *52*, 43–60. [CrossRef]
40. Salvador, A.; Veiga, J.P. A permanent signal related to male pairing success and survival in the lizard *Psammodromus algirus*. *Amphibia Reptilia* **2008**, *29*, 117–120.
41. Martín, J.; Forsman, A. Social costs and development of nuptial coloration in male *Psammodromus algirus* lizards: An experiment. *Behav. Ecol.* **1999**, *10*, 396–400. [CrossRef]
42. Díaz, J.A.; Iraeta, P.; Verdú-Ricoy, J.; Siliceo, I.; Salvador, A. Intraspecific variation of reproduction traits in a Mediterranean lizard: Clutch, population, and lineage effects. *Evol. Biol.* **2012**, *39*, 106–115. [CrossRef]
43. Brooks, R.C.; Kemp, D.J. Can older males deliver the good genes? *Trends Ecol. Evol.* **2001**, *16*, 308–313. [CrossRef]
44. Zamora-Camacho, F.J.; Reguera, S.; Moreno-Rueda, G.; Pleguezuelos, J.M. Patterns of seasonal activity in a Mediterranean lizard along a 2200m altitudinal gradient. *J. Thermal Biol.* **2013**, *38*, 64–69. [CrossRef]
45. Zamora-Camacho, F.J.; Reguera, S.; Moreno-Rueda, G. Bergmann's Rule rules body size in an ectotherm: Heat conservation in a lizard along a 2200-metre elevational gradient. *J. Evol. Biol.* **2014**, *27*, 2820–2828. [CrossRef] [PubMed]
46. Perry, G.; Wallace, M.C.; Perry, D.; Curzer, H.; Muhlberger, P. Toe clipping of amphibians and reptiles: Science, ethics, and the law. *J. Herpetol.* **2011**, *45*, 547–555. [CrossRef]
47. Comas, M.; Reguera, S.; Zamora-Camacho, F.J.; Salvadó, H.; Moreno-Rueda, G. Comparison of the effectiveness of phalanges vs. humeri and femurs to estimate lizard age with skeletochronology. *Anim. Biodiv. Conserv.* **2016**, *39*, 237–240. [CrossRef]
48. Zhao, M.; Klaassen, C.A.J.; Lisovski, S.; Klaassen, M. The adequacy of aging techniques in vertebrates for rapid estimation of population mortality rates from age distributions. *Ecol. Evol.* **2019**, *9*, 1394–1402. [CrossRef]
49. Andersson, S.; Prager, M. Quantifying colors. In *Bird Coloration. Vol. II: Function and Evolution*; Hill, G.E., McGraw, K.J., Eds.; Harvard University Press: Boston, MA, USA, 2006; pp. 41–89.
50. Montgomerie, R. Analazing colors. In *Bird Coloration. Vol. I: Mechanisms and Measurements*; Hill, G.E., McGraw, K.J., Eds.; Harvard University Press: Boston, MA, USA, 2006; pp. 90–147.
51. Endler, J.A. On the measurement and classification of colour in studies of animal colour patterns. *Biol. J. Linn. Soc.* **1990**, *41*, 315–352. [CrossRef]
52. Rasband, W.S. *ImageJ*; US National Institutes of Health: Bethesda, MD, USA, 2008.
53. Zuur, A.F.; Ieno, E.N.; Elphick, C.S. A protocol for data exploration to avoid common statistical problems. *Methods Ecol. Evol.* **2010**, *1*, 3–14. [CrossRef]
54. Packard, G.C. Is logarithmic transformation necessary in allometry? *Biol. J. Linn. Soc.* **2013**, *109*, 476–486. [CrossRef]
55. R Development Core Team. *R: A Language and Environment for Statistical Computing*; R Foundation for Statistical Computing: Vienna, Austria, 2019.
56. Ripley, B.; Venables, W. *nnet: Feed-Forward Neural Networks and Multinomial Log-Linear Models*; R Foundation for Statistical Computing: Vienna, Austria, 2020.
57. Bartoń, K. *MuMIn: Multi-Model Inference*; R Foundation for Statistical Computing: Vienna, Austria, 2020.
58. Quinn, G.P.; Keough, M.J. *Experimental Design and Data Analysis for Biologists*; Cambridge University Press: Cambridge, UK, 2002.
59. Reguera, S.; Zamora-Camacho, F.J.; Moreno-Rueda, G. The lizard *Psammodromus algirus* (Squamata: Lacertidae) is darker at high altitudes. *Biol. J. Linn. Soc.* **2014**, *112*, 132–141. [CrossRef]
60. Huyghe, K.; Vanhooydonck, B.; Scheers, H.; Molina-Borja, M.; van Damme, R. Morphology, performance and fighting capacity in male lizards, *Gallotia galloti*. *Funct. Ecol.* **2005**, *19*, 800–807. [CrossRef]
61. Stapley, J.; Whiting, M.J. Ultraviolet signals fighting ability in a lizard. *Biol. Lett.* **2006**, *2*, 169–172. [CrossRef]
62. Pérez i de Lanuza, G.; Carazo, P.; Font, E. Colours of quality: Structural (but not pigment) coloration informs about male quality in a polychromatic lizard. *Anim. Behav.* **2014**, *90*, 73–81. [CrossRef]
63. Hamilton, D.G.; Whiting, M.J.; Pryke, S.R. Fiery frills: Carotenoid-based coloration predicts contest success in frillneck lizards. *Behav. Ecol.* **2013**, *24*, 1138–1149. [CrossRef]
64. Møller, A.P. Social control of deception among status signalling house sparrows *Passer domesticus*. *Behav. Ecol. Sociobiol.* **1987**, *20*, 307–311. [CrossRef]

65. Lappin, A.K.; Brandt, Y.; Husak, J.F.; Macedonia, J.M.; Kemp, D.J. Gaping displays reveal and amplify a mechanically based index of weapon performance. *Am. Nat.* **2006**, *168*, 100–113. [CrossRef] [PubMed]
66. Weiss, S.L. Female-specific color is a signal of quality in the striped plateau lizard (*Sceloporus virgatus*). *Behav. Ecol.* **2006**, *17*, 726–732. [CrossRef]
67. Lande, R. Sexual dimorphism, sexual selection, and adaptation in polygenic characters. *Evolution* **1980**, *34*, 292–305. [CrossRef]
68. López, P.; Martín, J.; Cuadrado, M. Chemosensory cues allow male lizards *Psammodromus algirus* to override visual concealment of sexual identity by satellite males. *Behav. Ecol. Sociobiol.* **2003**, *54*, 218–224. [CrossRef]
69. Searcy, W.A.; Nowicki, S. *The Evolution of Animal Communication*; Princeton University Press: Princeton, NJ, USA, 2005.
70. Martín, J.; López, P. Nuptial coloration and mate guarding affect escape decision of male lizards *Psammodromus algirus*. *Ethology* **1999**, *105*, 439–447. [CrossRef]
71. Dunn, P.O.; Armenta, J.K.; Whittingham, L.A. Natural and sexual selection act on different axes of variation in avian plumage color. *Sci. Adv.* **2015**, *1*, e1400155. [CrossRef] [PubMed]
72. Comas, M.; Reguera, S.; Zamora-Camacho, F.J.; Moreno-Rueda, G. Age structure of a lizard along an elevational gradient reveals nonlinear lifespan patterns with altitude. *Curr. Zool.* **2020**, *142*, 373–382. [CrossRef] [PubMed]
73. Zamora-Camacho, F.J.; Reguera, S.; Moreno-Rueda, G. Thermoregulation in the lizard *Psammodromus algirus* along a 2200-m elevational gradient in Sierra Nevada (Spain). *Int. J. Biometeorol.* **2016**, *60*, 687–697. [CrossRef] [PubMed]
74. Olsson, M.; Wapstra, E.; Schwartz, T.; Madsen, T.; Ujvari, B.; Uller, T. In hot pursuit: Fluctuating mating system and sexual selection in san lizards. *Evolution* **2011**, *65*, 574–583. [CrossRef] [PubMed]
75. Václav, R.; Prokop, P.; Fekiač, V. Expression of breeding coloration in European Green Lizards (*Lacerta viridis*): Variation with morphology and tick infestation. *Can. J. Zool.* **2007**, *85*, 1199–1206. [CrossRef]
76. Calisi, R.M.; Malone, J.H.; Hews, D.K. Female secondary coloration in the Mexican boulder spiny lizard is associated with nematode load. *J. Zool.* **2008**, *276*, 358–367. [CrossRef]
77. Martín, J.; Amo, L.; López, P. Parasites and health affect multiple sexual signals in male common wall lizards, *Podarcis muralis*. *Naturwissenschaften* **2008**, *95*, 293–300. [CrossRef]
78. Megía-Palma, R.; Martínez, J.; Merino, S. A structural colour ornament correlates positively with parasite load and body condition in an insular lizard species. *Sci. Nat.* **2016**, *103*, 52. [CrossRef]
79. Megía-Palma, R.; Martínez, J.; Merino, S. Manipulation of parasite load induces significant changes in the structural-based throat color of male Iberian green lizards. *Curr. Zool.* **2018**, *64*, 293–302. [CrossRef]
80. Megía-Palma, R.; Paranjpe, D.; Reguera, S.; Martínez, J.; Cooper, R.D.; Blaimont, P.; Merino, S.; Sinervo, B. Multiple color patches and parasites in *Sceloporus occidentalis*: Differential relationships by sex and infection. *Curr. Zool.* **2018**, *64*, 703–711. [CrossRef]
81. Álvarez-Ruiz, L.; Megía-Palma, R.; Reguera, S.; Ruiz, S.; Zamora-Camacho, F.J.; Figuerola, J.; Moreno-Rueda, G. Opposed elevational variation in prevalence and intensity of endoparasites and their vectors in a lizard. *Curr. Zool.* **2018**, *64*, 197–204. [CrossRef]
82. Megía-Palma, R.; Martínez, J.; Merino, S. Structural- and carotenoid-based throat colour patches in males of *Lacerta schreiberi* reflect different parasitic diseases. *Behav. Ecol. Sociobiol.* **2016**, *70*, 2017–2025.
83. Moreno-Rueda, G.; Melero, E.; Reguera, S.; Zamora-Camacho, F.J.; Álvarez-Benito, I. Prey availability, prey selection, and trophic niche width in the lizard *Psammodromus algirus* along an elevational gradient. *Curr. Zool.* **2018**, *64*, 603–613. [CrossRef] [PubMed]
84. Cuervo, J.J.; Belliure, J.; Negro, J.J. Coloration reflects skin pterin concentration in a red-tailed lizard. *Comp. Biochem. Physiol. B* **2016**, *193*, 17–24. [CrossRef]
85. Hill, G.E. Geographic variation in the carotenoid plumage pigmentation of male house finches (*Carpodacus mexicanus*). *Biol. J. Linn. Soc.* **1993**, *49*, 63–86. [CrossRef]
86. Grether, G.F.; Hudon, J.; Millie, D.F. Carotenoid limitation of sexual coloration along an environmental gradient in guppies. *Proc. R. Soc. B* **1999**, *266*, 1317–1322. [CrossRef]
87. Reguera, S.; Zamora-Camacho, F.J.; Trenzado, C.E.; Sanz, A.; Moreno-Rueda, G. Oxidative stress decreases with elevation in the lizard *Psammodromus algirus*. *Comp. Biochem. Physiol. A* **2014**, *172*, 52–56. [CrossRef]
88. Reguera, S.; Zamora-Camacho, F.J.; Melero, E.; García-Mesa, S.; Trenzado, C.E.; Cabrerizo, M.J.; Sanz, A.; Moreno-Rueda, G. Ultraviolet radiation does not increase oxidative stress in the lizard *Psammodromus algirus* along an elevational gradient. *Comp. Biochem. Physiol. A* **2015**, *183*, 20–26. [CrossRef]
89. McGraw, K.J. The antioxidant function of many animal pigments: Are there consistent health benefits of sexually selected colourants? *Anim. Behav.* **2005**, *69*, 757–764. [CrossRef]
90. Olsson, M.; Tobler, M.; Healey, M.; Perrin, C.; Wilson, M. A significant component of ageing (DNA damage) is reflected in fading breeding colors: An experimental test using innate antioxidant mimetics in painted dragon lizards. *Evolution* **2012**, *66*, 2475–2483. [CrossRef]
91. Jennions, M.D.; Petrie, M. Variation in mate choice and mating preferences: A review of causes and consequences. *Biol. Rev.* **1997**, *72*, 283–327. [CrossRef]
92. Brooks, R.C.; Couldridge, V. Multiple sexual ornaments coevolve with multiple mating preferences. *Am. Nat.* **1999**, *154*, 37–45. [CrossRef]
93. Kwiatkowski, M.A.; Sullivan, B.K. Geographic variation in sexual selection among populations of an iguanid lizard, *Sauromalus obesus* (=*ater*). *Evolution* **2002**, *56*, 2039–2051. [CrossRef] [PubMed]

94. Endler, J.A. The color of light in forests and its implications. *Ecol. Monogr.* **1993**, *63*, 1–27. [CrossRef]
95. Díaz, J.A.; Verdú-Ricoy, J.; Iraeta, P.; Llanos-Garrido, A.; Pérez-Rodríguez, A.; Salvador, A. There is more to the picture than meets the eye: Adaptation for crypsis blurs phylogeographical structure in a lizard. *J. Biogeogr.* **2017**, *44*, 397–408. [CrossRef]
96. Stuart-Fox, D.M.; Moussalli, A.; Marshall, N.J.; Owens, I.P.F. Conspicuous males suffer higher predation risk: Visual modelling and experimental evidence from lizards. *Anim. Behav.* **2003**, *66*, 541–550. [CrossRef]

Article

Effects of Caudal Autotomy on the Locomotor Performance of *Micrablepharus Atticolus* (Squamata, Gymnophthalmidae)

Naiane Arantes Silva [1], Gabriel Henrique de Oliveira Caetano [2], Pedro Henrique Campelo [3], Vitor Hugo Gomes Lacerda Cavalcante [4], Leandro Braga Godinho [1], Donald Bailey Miles [5], Henrique Monteiro Paulino [3], Júlio Miguel Alvarenga da Silva [1], Bruno Araújo de Souza [1], Hosmano Batista Ferreira da Silva [1] and Guarino Rinaldi Colli [3,*]

1 Programa de Pós-Graduação em Ecologia e Conservação, Campus Nova Xavantina, Universidade do Estado de Mato Grosso, Rua Prof. Dr. Renato Figueiro Varella, Nova Xavantina 78690, MT, Brazil; naianearantes.bio@gmail.com (N.A.S.); lbgcarranca@gmail.com (L.B.G.); julio7alvarenga@gmail.com (J.M.A.d.S.); souza_bruno@icloud.com (B.A.d.S.); hosmanobatista@gmail.com (H.B.F.d.S.)

2 Jacob Blaustein Center for Scientific Cooperation, The Jacob Blaustein Institutes for Desert Research, Ben-Gurion University of the Negev, Midreshet Ben-Gurion 849900, Israel; gabrielhoc@gmail.com

3 Departamento de Zoologia, Campus Universitário Darcy Ribeiro, Universidade de Brasília, Asa Norte, Brasília 70910, DF, Brazil; pedro.h.campelo@gmail.com (P.H.C.); hunterh1008@gmail.com (H.M.P.)

4 Instituto Federal do Piauí, Teresina 64000-040, Piauí, Brazil; vitor.cavalcante@ifpi.edu.br

5 Department of Biology, Ohio University, Athens, OH 45701, USA; urosaurus@gmail.com

* Correspondence: grcolli@unb.br

Citation: Silva, N.A.; Caetano, G.H.d.O.; Campelo, P.H.; Cavalcante, V.H.G.L.; Godinho, L.B.; Miles, D.B.; Paulino, H.M.; da Silva, J.M.A.; de Souza, B.A.; da Silva, H.B.F.; et al. Effects of Caudal Autotomy on the Locomotor Performance of *Micrablepharus Atticolus* (Squamata, Gymnophthalmidae). *Diversity* **2021**, *13*, 562. https://doi.org/10.3390/d13110562

Academic Editors: Michael Wink and Francisco Javier Zamora-Camacho

Received: 6 July 2021
Accepted: 9 September 2021
Published: 4 November 2021

Abstract: Caudal autotomy is a striking adaptation used by many lizard species to evade predators. Most studies to date indicate that caudal autotomy impairs lizard locomotor performance. Surprisingly, some species bearing the longest tails show negligible impacts of caudal autotomy on sprint speed. Part of this variation has been attributed to lineage effects. For the first time, we model the effects of caudal autotomy on the locomotor performance of a gymnophthalmid lizard, *Micrablepharus atticolus*, which has a long and bright blue tail. To improve model accuracy, we incorporated the effects of several covariates. We found that body temperature, pregnancy, mass, collection site, and the length of the regenerated portion of the tail were the most important predictors of locomotor performance. However, sprint speed was unaffected by tail loss. Apparently, the long tail of *M. atticolus* is more useful when using undulation amidst the leaf litter and not when using quadrupedal locomotion on a flat surface. Our findings highlight the intricate relationships among physiological, morphological, and behavioral traits. We suggest that future studies about the impacts of caudal autotomy among long-tailed lizards should consider the role of different microhabitats/substrates on locomotor performance, using laboratory conditions that closely mimic their natural environments.

Keywords: lizard; autotomy; tail; locomotion; performance; temperature; predation

1. Introduction

Throughout evolutionary time, an "arms race" fostered varied strategies of prey capture and predator escape [1]. Autotomy—the self-amputation of a body part in response to an attack by a predator—is one of the most dramatic adaptations to avoid predation [2]. Caudal autotomy among reptiles has an ancient origin and was present in captorhinids from the Early Permian [3]. It persists to this day among squamate reptiles, in some species of snakes and most lizards, allowing them to escape while the predator is distracted by the abandoned tail part [4–7]. The detachment of the tail in most species occurs through pre-established, intravertebral fracture planes, the oldest and most common form of autotomy to date, allowing a new tail to grow supported by a calcified cartilage tube [8–11].

Despite the immediate benefit of avoiding predation, autotomy also involves energy costs that can influence survival. For instance, even when resources are limiting, tail regeneration is a priority, probably associated with long-term survival and reproductive success [12,13]. Thus, the production of a new tail can negatively affect energy balance, immunity, growth rate, social *status*, and immediate reproductive success [2,14]. Besides, autotomy results in the temporary loss of an important mechanism to avoid predation. Therefore, autotomized individuals may alter their patterns of activity and space use, as well as foraging schedules and frequencies, to minimize exposure to predators [15,16].

In addition to affecting behavior, caudal autotomy affects the locomotor performance of some species. The tail is a counterweight, balancing the head and body during racing; therefore, its absence results in weight transfer to the forelimbs, making it challenging to move [4]. Moreover, the tail can act as an inertial damper of pelvic girdle movements, and its loss causes disordered oscillation of the hind limbs during the race [5] and reduced jump stability and performance [17]. In general, caudal autotomy leads to decreased locomotor performance [18]. However, it may not interfere [19,20] or even increase locomotor performance [21]. These opposite results may relate to interspecific differences in predation intensity throughout ontogeny, life habits, and sexual dimorphism [15]. For example, in sexually dimorphic species where males have conspicuous coloration, their locomotor performance is little affected by autotomy, as potential predators and competitors can easily see them [18,22]. Still, variation exists between and within evolutionary lineages associated with different tail shapes and functions, such as sexual displays, predator distraction, defense, balance, fat storage, stabilization, and an auxiliary organ in climbing [20,23].

The lizard genus *Micrablepharus* (Squamata, Gymnophthalmidae) contains two species: *M. maximiliani* (Reinhardt and Lütken, 1861), widely distributed across the South American dry diagonal, comprising the Chaco, Cerrado, and Caatinga, and *M. atticolus* Rodrigues, 1996, endemic to the Cerrado [24–27]. The two species are diurnal, semifossorial, and live among the leaf litter [28–31]. Reproductive activity peaks in the dry season, and populations undergo an almost complete annual replacement [32,33]. They share an elongate trunk and tail, short limbs, and digit reduction on the forelimbs (complete loss of digit I), whereas the hindlimbs follow the pentadactyl condition [34,35]. They exhibit intermittent quadrupedal locomotion, combining conspicuous axial traveling waves with trot-like coordination of the limbs [36–38]. The vertebral axis is the main effector of locomotion, while the limbs play an auxiliary role. On low friction substrates, the axial system of locomotion predominates, but the limbs become increasingly involved as substrate friction increases and with increasing speed [37]. *Micrablepharus atticolus* and *M. maximiliani* have a long and bright blue tail that contributes to divert attention from visually oriented predators to a non-vital part of the body at the time of an attack, which may be associated with higher rates of autotomy in more open environments [39].

Tail loss in *Micrablepharus atticolus* does not affect body condition, suggesting that the energetic costs of autotomy are low or that individuals compensate for the tail loss by increasing foraging rate [39]. Consequently, autotomy may not impair locomotor performance by reducing energy reserves [40]. However, because of the importance of the axial system during locomotion [37], tail loss may compromise sprint speed. Locomotor performance is an essential determinant of fitness, because its reduction can undermine survival, reproductive success [41,42], foraging [43,44], and social dominance [45]. Since environmental variation affects autotomy rates in *M. atticolus*, but these do not affect survival [39], studying the effect of autotomy on the locomotor performance of lizards inhabiting different environments can contribute to the understanding of possible compensatory mechanisms.

Here, we investigate the effects of caudal autotomy on the locomotor performance of *Micrablepharus atticolus* from two different environments, one in the central Cerrado and another in the Cerrado-Amazonia transition. We take into account the effects of geography, sex, body temperature, and ontogeny since (1) locomotor performance tends to be lower in females, especially during pregnancy, by the effect of the additional burden represented by the litter [46,47]; (2) there is a positive allometric relationship between body size and locomotor performance [48]; and (3) central Cerrado lizards are expected to have better locomotor performance, assuming that environmental conditions should be optimal for performance in the core of species distributions [49]. Moreover, we investigate whether the effects of autotomy on locomotor performance are proportional to the size of the remaining or regenerated portion of the tail [50].

2. Materials and Methods

2.1. Study Sites

We collected data from lizards captured at two sites: Reserva do IBGE (15°56′06″ S, 47°52′09″ W), a protected area in Brasília, Distrito Federal, Brazil, in the central area of the Cerrado; and Parque do Bacaba (14°42′24″ S, 52°21′10″ W), Nova Xavantina, Mato Grosso, Brazil, in the Cerrado-Amazon transition. The climate in both sites is tropical with dry winter, Aw in Köppen's classification [51], with a dry season from May to September and a rainy season from October to April. In Brasília, the average annual accumulated precipitation is 1477.4 mm, and the average annual temperature is 21.0 °C; in Nova Xavantina, 1417.7 mm and 24.8 °C, respectively (https://portal.inmet.gov.br/normais, accessed on 30 March 2021).

2.2. Lizard Sampling

We captured lizards using arrays of pitfall traps interconnected by drift fences, as part of a long-term, mark-recapture study on their demography and community dynamics. Each array consisted of four plastic buckets of 35 L, buried to ground level and arranged in the form of a "Y", interconnected by three 6 m long and 50 cm high galvanized steel plates that functioned as guide fences. Immediately after capture, we took the following measurements from each lizard: body mass, using a Pesola spring dynamometer (0.1 g precision); snout-vent length (SVL), total tail length, and length of the non-autotomized part of the tail—in lizards with caudal autotomy—with a metal ruler (1 mm precision); and sex, whenever possible, through palpation of the abdomen for the presence of vitellogenic follicles or eggs in pregnant females and the extrusion of the hemipenis in males. Next, we transported lizards to the lab and housed them in individual terraria, with vermiculite substrate and water ad libitum. Up to 24 h after capture, we carried out ecophysiology experiments (below), after which we permanently marked (by toe-clipping) and released lizards next to their exact capture sites. We captured and handled all individuals with great care to prevent any damage to the tail, such that autotomized tails resulted exclusively from natural processes. Finally, we only used adult individuals in the analyses, comprising 39 lizards from Brasília and 64 from Nova Xavantina. We considered individuals with SVL greater than 35 mm as adults [32].

2.3. Locomotor Performance

We recorded sprint speed on a wooden track (300 cm long × 30 cm high × 40 cm wide). We induced each lizard to run as fast as possible by manual stimulation, mimicking a predatory chase, to record the maximum speed. Due to the thermal sensitivity of sprint speed [52], we conducted runs at three different temperatures—cold (=ambient −5 °C), ambient (~20 °C), and hot (=ambient +5 °C)—in each experiment. We used gel ice packs and incandescent lamps to alter lizards' body temperature, monitored with a fast-reading cloacal thermometer (L-K Industries Miller & Weber T-6000 Cloacal 0/50 °C 0.2 precision). We conducted two trials of each lizard at each temperature, totaling six runs. We recorded runs at 420 fps with a Casio HS EX-FH25 digital camera mounted on an aluminum tripod at 1.5 m height in the center of the track. Later, we analyzed videos with Tracker 4.80 to obtain the maximum sprint speed of each lizard at each temperature.

Within at least one hour after the last run, we measured the critical thermal minimum and maximum, with a one-hour interval between them, using a fast-reading cloacal thermometer (L-K Industries Miller & Weber T-6000 Cloacal 0/50 °C 0.2 precision). We exposed lizards to the sources of heat and cold mentioned above until they lost the righting response, i.e., when they could not return to the prone position after turning in a supine position without leading the animal to death. To build performance curves (below), we considered that sprint speed is equal to zero at the critical thermal minimum and maximum. The Animal Use Ethics Committee of the University of Brasília approved all procedures (process 33786/2016).

2.4. Statistical Analyses

To determine the effect of caudal autotomy on locomotor performance, we built generalized mixed-effects additive models—GAMMs [53] with the MGCV package [54] in the R environment [55]. We used GAMMs to generate performance curves and evaluate the influence of predictors on the shape and location of curves because they allow nonlinear responses and are flexible due to the non-parametric smoothing functions used in sections of the data [56]. In these models, we used the maximum sprint speed as the response variable; the individual as a random factor; and sex, pregnancy (gravid/non-gravid), body temperature, mass, SVL, relative tail length (total tail length/SVL), caudal autotomy (autotomized/not autotomized), relative length of the regenerated portion of the tail (length of the regenerated portion of tail/SVL), and study site (Brasília/Nova Xavantina) as fixed factors. To assess model significance, we used a likelihood-ratio test comparing its fit with that of a null model, composed only of the response variable, the intercept, and the random factor.

To evaluate predictor importance, we used a combination of model selection and averaging based on the Akaike Information Criterion adjusted for small samples (AIC_c), with the package MuMIn [57]. Model selection attempts to improve our understanding of the relationship between the response and the predictors by reducing model's complexity. However, this approach often results in biased regression parameters and too small respective standard errors in finite samples because they do not reflect the uncertainty related to the model selection process [58,59]. On the other hand, model averaging incorporates the uncertainty intrinsic to model selection by combining parameter estimates across different models [60,61]. Using this approach, we examined the complete set of possible models combining the fixed effects to obtain model-averaged standardized parameter estimates for statistical inference [62,63]. We used averages calculated across all models ("full averages"), assuming that each model includes all variables, but that in some models the corresponding coefficient (and its respective variance) is set to zero, which avoids biasing the values away from zero [64]. Moreover, we calculated the importance of each predictor as the sum of Akaike weights across all models containing that predictor.

3. Results

We obtained ecophysiological data from 39 lizards from Brasília and 64 lizards from Nova Xavantina (Table 1). The likelihood-ratio test indicated that our full GAMM differed significantly from a null model and adequately fitted the data ($\chi^2_{[1]} = 219.129$, $p < 0.001$, adjusted-$r^2 = 0.622$). Among the parametric terms in the model, collection site and pregnancy were significant, while among smooth terms, body temperature, body mass, and the length of the regenerated portion of the tail were significant (Table 2). Model selection and averaging indicated that body temperature, pregnancy, mass, collection site, and the length of the regenerated portion of the tail, in this order, were the most important predictors of locomotor performance in *Micrablepharus atticolus* (Table 3). The GAMM predicted maximum locomotor performance around 31 °C (Figure 1A). Gravid females had lower performance than males and non-gravid females (Figure 1B), and lizards from Nova Xavantina achieved higher performance—and at higher temperatures—than lizards from Brasília (Figure 1C). Finally, the locomotor performance increased with body mass (Figure 2A) and the relative length of the regenerated portion of the tail (Figure 2B).

Table 1. Summary statistics of ecophysiological parameters of *Micrablepharus atticolus* from Brasília and Nova Xavantina, Brazil. Values represent the mean ± one standard deviation.

Parameter	Brasília	Nova Xavantina	Total
Sample size	39	64	103
Mass (g)	1.49 ± 0.35	0.82 ± 0.19	1.06 ± 0.41
Snout-vent length (mm)	38.22 ± 3.06	34.98 ± 2.76	36.12 ± 3.26
Tail length (mm)	48.62 ± 13.74	46.37 ± 17.64	47.17 ± 16.39
Relative tail length	1.27 ± 0.34	1.32 ± 0.50	1.31 ± 0.45
Length of regenerated portion of tail (mm)	6.38 ± 10.30	9.99 ± 11.95	8.71 ± 11.51
Relative length of regenerated portion of tail	0.16 ± 0.26	0.28 ± 0.34	0.24 ± 0.32
Body temperature during runs (°C)	27.94 ± 6.72	27.53 ± 5.43	27.67 ± 5.88
Critical thermal minimum (°C)	13.90 ± 1.78	15.20 ± 3.29	14.70 ± 2.08
Critical thermal maximum (°C)	44.41 ± 1.84	40.00 ± 3.09	41.69 ± 3.43
Sprint speed (maximum)	0.07 ± 0.02	0.10 ± 0.04	0.09 ± 0.04

Table 2. Full generalized additive mixed-effects model (GAMM) relating predictors to locomotor performance (sprint speed) of the lizard *Micrablepharus atticolus*. AU: caudal autotomy (yes/no), CCr: total tail length, RCr: length of the regenerated portion of the tail, SVL: snout-vent length, NX: Nova Xavantina, edf: expected degrees-of-freedom.

Parametric Terms				
Term	**Estimate**	**Std. Error**	***t***	***p***
(Intercept)	0.0411	0.0065	6.3080	<0.0001
LocalNX	0.0220	0.0069	3.1920	0.0015
SexMale	0.0015	0.0045	0.3230	0.7467
AUYes	0.0000	0.0070	−0.0040	0.9970
Pregnancy	−0.0203	0.0068	−2.9700	0.0031
Smooth Terms				
Term	**edf**	**Ref. df**	***F***	***p***
s(Temperature)	7.976	7.976	95.569	<0.00001
s(CCr)	1.000	1.000	0.036	0.84979
s(RCr)	1.000	1.000	3.037	0.08203
s(SVL)	1.000	1.000	0.365	0.54589
s(Mass)	3.119	3.119	4.983	0.00259

Table 3. Model selection and averaging of generalized additive mixed-effects models (GAMMs) relating predictors to locomotor performance (sprint speed) of the lizard *Micrablepharus atticolus*. Models depicted are those with $\Delta AIC_c < 4$. AU: caudal autotomy (yes/no), RT: length of regenerated portion of tail, AIC_c: Akaike information criterion corrected for small samples, ΔAIC_c: difference between given and best model, $wAIC_c$: Akaike weight.

Model Selection					
Model	**df**	**logLik**	**AICc**	**ΔAICc**	**wAICc**
Pregnancy + Site + s(Mass) + s(RCr) + s(Temperature)	11	934.81	−1847.06	0.00	0.25
AU + Pregnancy + Site + s(Mass) + s(RCr) + s(Temperature)	12	934.84	−1845.01	2.05	0.09
Pregnancy + Site + s(Mass) + s(RCr) + s(Temperature)+ Sex	12	934.82	−1844.97	2.10	0.09
Pregnancy + Site + s(Mass) + s(Temperature)	9	931.35	−1844.32	2.75	0.06
AU + Pregnancy + Site + s(Mass) + s(Temperature)	10	932.22	−1843.96	3.10	0.05
Pregnancy + Site + s(Mass) + s(RCr) + s(SVL) + s(Temperature)	13	934.93	−1843.07	3.99	0.03

Model Averaging									
Importance	**s(Temperature)**	**Pregnancy**	**s(Mass)**	**Site**	**s(RCr)**	**AU**	**Sex**	**s(CCr)**	**s(SVL)**
Sum of model weights	1.00	0.93	0.90	0.89	0.73	0.33	0.27	0.15	0.14
Number of containing models	255	253	253	254	254	255	255	255	254

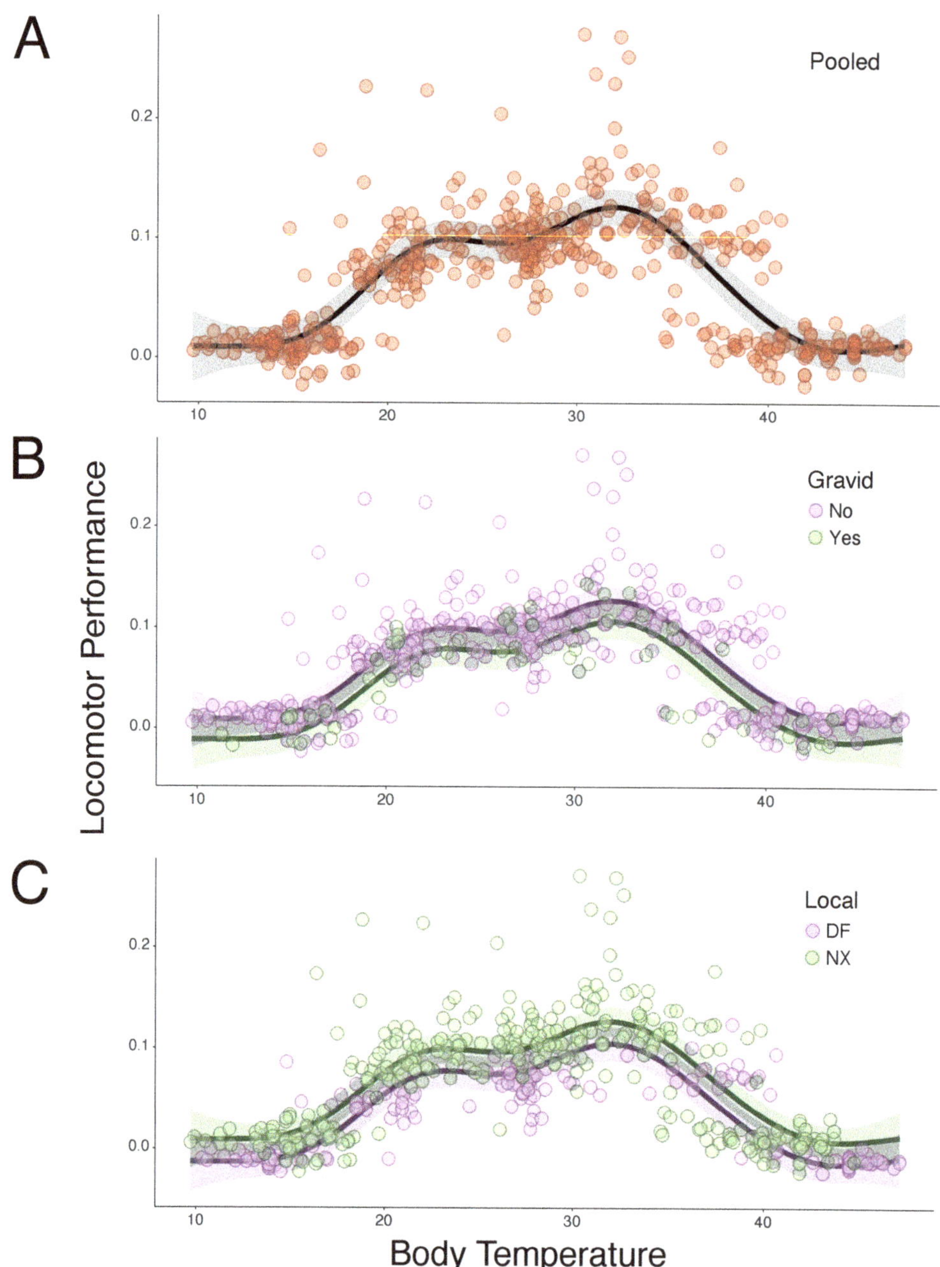

Figure 1. Locomotor performance (sprint speed) of the lizard *Micrablepharus atticolus* as a function of (**A**) body temperature, (**B**) body temperature and female reproductive condition (gravid females vs. non-gravid females and males), and (**C**) body temperature and geography. Points represent partial residuals of a generalized additive mixed model (GAMM), while lines and bands represent the predictions and confidence limits, respectively.

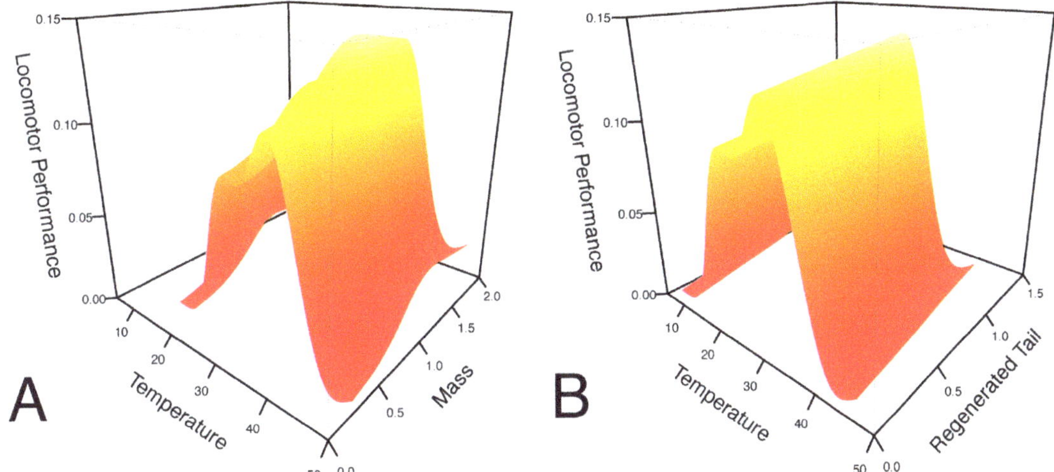

Figure 2. Locomotor performance (sprint speed) of the lizard *Micrablepharus atticolus* as a function of (**A**) body temperature and body mass, and (**B**) body temperature and length of the regenerated portion of the tail. The surface represents the predictions of a generalized additive mixed model (GAMM).

4. Discussion

We assessed the effects of caudal autotomy on the locomotor performance of *Micrablepharus atticolus*, controlling for the influence of several covariates. We found that the performance is significantly affected by body temperature, female reproductive condition, body mass, geography, and caudal autotomy. Overall, our findings highlight the complex patterns of association among physiological, morphological, and behavioral traits and that meaningful inference and prediction based on physiological performance must consider such patterns [65–68].

Body temperature was the foremost factor affecting performance. This outcome is not surprising, given that body temperature is one of the most critical ecophysiological variables affecting the performance of ectotherms [69–71]. Sprint speed peaked at ca. 31 °C, which is substantially higher than that recorded for *Caparaonia itaiquara* (24.51 °C) and *Colobodactylus dalcianus* (25.81 °C), two closely related gymnophthalmines from high-elevation areas in the Atlantic Forest of southeastern Brazil [72,73]. Moreover, our analyses showed that lizards from Nova Xavantina achieve higher sprint speeds at higher body temperatures than lizards from Brasília. Such differences might be related to altitudinal, latitudinal, or even lineage effects [74]. As environmental temperatures in Nova Xavantina are ca. 4 °C higher than in Brasília (and even higher than in high elevations of southeastern Brazil), our results are consistent with the notion that geographic variation of thermal sensitivity in locomotor performance is adaptive, such that organisms adjust optimal performance temperatures to prevalent field body temperatures [67]. For instance, based on the principle that biochemical and physiological systems operating at high temperatures have a high catalytic capacity, the "hotter is better" hypothesis predicts a positive relationship between maximal organismal performance and optimal temperatures [75]. This relationship holds when considering interspecific [65,76] or intraspecific comparisons [77,78].

Whereas Brasília is at the core of *Micrablepharus atticolus*' geographic distribution, Nova Xavantina is closer to its periphery, next to the Cerrado–Amazonia ecotone [26,79]. Therefore, we expected higher physiological performance in the core population, as predicted by the core-periphery hypothesis [49,80]. However, we found the opposite pattern, with higher performance in the more peripheral population. Several factors might account for this result. For example, despite the centrality difference between the two sites relative to the species' range, they might have the same or even opposite patterns of environmental suitability, i.e., the geometric center of the geographic distribution may not coincide with areas of greater suitability and vice-versa [81,82]. Moreover, due to phenotypic plasticity or adaptation to local conditions, species range boundaries may not be driven by thermal performance [72,83].

Our analysis indicates no intersexual differences in sprint speed in *Micrablepharus atticolus*, most likely related to the lack of sexual size dimorphism [29]. However, we found that pregnant females have lower locomotor performance than males. This decrease likely occurs due to the additional physical load of the litter, making the body broader and heavier [84]. However, as locomotor performance increased with body mass, the lower performance in pregnant females may be related to physiological changes linked to reproduction [85], such as decreased muscle strength, reduced metabolic capacity, motivation to escape [86], and energy allocation [87]. These physiological changes ensure adequate embryonic development and remain for a while after egg-laying [85,88]. A decrease in gravid females' locomotor performance was also recorded in other lizard species [89–93]. By becoming slower, pregnant females are more susceptible to predation, and this can promote several behavioral changes during pregnancy, such as foraging near potential shelters and avoiding long races during a predatory escape.

The body mass of individuals is an essential factor in determining sprint speed [76]. We found continuously increased performance with increasing body mass, which would probably occur until the optimal mass is reached, beyond which performance decreases [94,95]. This increase in performance with body mass is typical among quadruped species [45,76,96,97]. Despite using lateral undulation when moving in the middle of the leaf litter, *Micrablepharus atticolus* can also rely on quadrupedal locomotion when on a flat substrate [36,37]. We advance that the ever-increasing locomotor performance associated with increased body mass results from the very short lifespan of *M. atticolus* [32], such that individuals never reach a critical body mass.

In most cases, the tail has an active role in improving lizard sprint speed, and caudal autotomy undermines locomotor performance [14,20]. Moreover, the greater the relative size of the intact tail, the higher the magnitude of sprint speed change following autotomy. However, we found that sprint speed was unaffected by tail loss but by the relative length of the regenerated portion of the tail, i.e., the longer the regenerated tail, the higher the sprint speed. Still, this effect was meager, unlike patterns documented elsewhere for eublepharids, lacertids, and skinks, [46,98–100]. Caudal autotomy has no impact on the locomotor performance of some lizard species [19,20]. Some researchers have argued that this reflects these species' skinny and short tails [19] or even that adverse effects of autotomy result from researchers damaging the lizards' locomotor muscles during experimental tail breakage [101]. Individuals of *Micrablepharus atticolus* have a long tail (in our samples, ~1.7× SVL in individuals with intact tails), one of the longest among gymnophthalmids [102], and we used lizards with naturally broken and regenerated tails. Therefore, these explanations cannot account for the patterns we observed.

A synthesis on the effects of tail autotomy, tail size, and locomotor performance in lizards identified clear phylogenetic patterns in the data [20]. Hence, among-lineage differences in the biomechanics of locomotion and the tail function during sprinting may account for the different effects of tail loss on locomotor performance. The single previous study on the locomotion of *Micrablepharus* did not address the impact of caudal autotomy on performance [37], and to the best of our knowledge, ours is the first study on this issue within Gymnophthalmidae. This lineage comprises small, cryptic, and often fossorial or semifossorial Neotropical species, characterized by many instances of the evolution of body elongation and limb reduction [34,35,103]. Indeed, fossoriality is a critical driver of the evolution of a snake-like morphology among squamates [104,105]. Therefore, in such species, the tail may have a very context-specific role in locomotion, which may not be apparent when individuals move on a flat substrate. For instance, in *Colobodactylus taunayi*, a gymnophthalmine, the tail remains stretched during displacement on a flat surface [102] and a similar pattern is apparent in *M. maximiliani* when moving on gravel or sand (Figures 3 and 6 in [37]). Tail loss in lizards of the genus *Takydromus*, where the tail can be three times as long as the SVL, similarly had little effect on locomotor performance [20,50,106]. These species often use a three-dimensional, cluttered environment amidst the leaf litter, much like "grass-swimmer" lizards [107,108]. We conjecture that the long tail of *M. atticolus* is more useful when using undulation amidst the leaf litter and not when using quadrupedal locomotion on a flat surface. Future studies on the impacts of caudal autotomy on long-tailed lizards should consider the role of different microhabitats/substrates on locomotor performance, using laboratory conditions that closely mimic their natural environment.

Author Contributions: N.A.S., H.M.P. and G.R.C. conceived the study, conducted the analyses, and wrote the manuscript. All authors participated in fieldwork, lab work, data collection and curation, and revised the manuscript. D.B.M., G.H.d.O.C. and G.R.C. provided planning and supervision. All authors have read and agreed to the published version of the manuscript.

Funding: GRC was supported by Coordenação de Aperfeiçoamento de Pessoal de Nível Superior (CAPES), Conselho Nacional de Desenvolvimento Científico e Tecnológico (CNPq), Fundação de Apoio à Pesquisa do Distrito Federal (FAPDF) and the USAID's PEER program under cooperative agreement AID-OAA-A-11-00012 for financial support. DBM was supported by NSF (1241848, 1950636).

Institutional Review Board Statement: The study was conducted according to the guidelines of the Declaration of Helsinki, and approved by the Institutional Ethics Committee of the University of Brasília (protocol code UnBDoc33786/2016 from 11 April 2016).

Informed Consent Statement: The study was conducted according to the guidelines of the Declaration of Helsinki, and approved by the Institutional Ethics Committee of the University of Brasília (protocol code UnBDoc33786/2016 from 11 April 2016).

Data Availability Statement: The data presented in this study are openly available in Dryad at https://doi.org/10.5061/dryad.gxd2547nc.

Conflicts of Interest: The authors declare no conflict of interest.

References

1. Dawkins, R.; Krebs, J.R. Arms races between and within species. *Proc. R. Soc. Lond. B Biol. Sci.* **1979**, *205*, 489–511. [PubMed]
2. Emberts, Z.; Escalante, I.; Bateman, P.W. The ecology and evolution of autotomy. *Biol. Rev.* **2019**, *94*, 1881–1896. [CrossRef]
3. LeBlanc, A.R.H.; MacDougall, M.J.; Haridy, Y.; Scott, D.; Reisz, R.R. Caudal autotomy as anti-predatory behaviour in Palaeozoic reptiles. *Sci. Rep.* **2018**, *8*, 3328. [CrossRef]
4. Arnold, E.N. Caudal autotomy as a defense. In *Biology of the Reptilia. Ecology B: Defense and Life History*; Gans, C., Huey, R.B., Eds.; Alan R. Liss, Inc.: New York, NY, USA, 1988; Volume 16.
5. Arnold, E.N. Evolutionary aspects of tail shedding in lizards and their relatives. *J. Nat. Hist.* **1984**, *18*, 127–169. [CrossRef]
6. Clause, A.R.; Capaldi, E.A. Caudal autotomy and regeneration in lizards. *J. Exp. Zool. Part A* **2006**, *305*, 965–973. [CrossRef]
7. Bellairs, A.d.A.; Bryant, S.V. Autotomy and regeneration in reptiles. In *Biology of the Reptilia, Volume 15, Development B*; Gans, C., Billet, F., Eds.; John Wiley & Sons: New York, NY, USA, 1985; pp. 301–410.
8. Gilbert, E.A.B.; Payne, S.L.; Vickaryous, M.K. The anatomy and histology of caudal autotomy and regeneration in lizards. *Physiol. Biochem. Zool.* **2013**, *86*, 631–644. [CrossRef] [PubMed]
9. Ritzman, T.B.; Stroik, L.K.; Julik, E.; Hutchins, E.D.; Lasku, E.; Denardo, D.F.; Wilson-Rawls, J.; Rawls, J.A.; Kusumi, K.; Fisher, R.E. The gross anatomy of the original and regenerated tail in the green anole (*Anolis carolinensis*). *Anat. Rec.* **2012**, *295*, 1596–1608. [CrossRef]
10. Alibardi, L. Development of the axial cartilaginous skeleton in the regenerating tail of lizards. *Bull. Assoc. Anat.* **1995**, *79*, 3–9.
11. Fisher, R.E.; Geiger, L.A.; Stroik, L.K.; Hutchins, E.D.; George, R.M.; Denardo, D.F.; Kusumi, K.; Rawls, J.A.; Wilson-Rawls, J. A histological comparison of the original and regenerated tail in the green anole, *Anolis carolinensis*. *Anat. Rec.* **2012**, *295*, 1609–1619. [CrossRef]
12. Maginnis, T.L. The costs of autotomy and regeneration in animals: A review and framework for future research. *Behav. Ecol.* **2006**, *17*, 857–872. [CrossRef]
13. Lynn, S.E.; Borkovic, B.P.; Russell, A.P. Relative apportioning of resources to the body and regenerating tail in juvenile leopard geckos (*Eublepharis macularius*) maintained on different dietary rations. *Physiol. Biochem. Zool.* **2013**, *86*, 659–668. [CrossRef]
14. Bateman, P.W.; Fleming, P.A. To cut a long tail short: A review of lizard caudal autotomy studies carried out over the last 20 years. *J. Zool.* **2009**, *277*, 1–14. [CrossRef]

15. Cooper, W.E., Jr.; Smith, C.S. Costs and economy of autotomy for tail movement and running speed in the skink *Trachylepis maculilabris*. *Can. J. Zool.* **2009**, *87*, 400–406. [CrossRef]
16. Cromie, G.L.; Chapple, D.G. Impact of tail loss on the behaviour and locomotor performance of two sympatric *Lampropholis* skink species. *PLoS ONE* **2012**, *7*, e34732. [CrossRef] [PubMed]
17. Gillis, G.B.; Bonvini, L.A.; Irschick, D.J. Losing stability: Tail loss and jumping in the arboreal lizard *Anolis carolinensis*. *J. Exp. Biol.* **2009**, *212*, 604–609. [CrossRef] [PubMed]
18. Anderson, M.L.; Cavalieri, C.N.; Rodríguez-Romero, F.; Fox, S.F. The differential effect of tail autotomy on sprint performance between the sexes in the lizard *Uta stansburiana*. *J. Herpetol.* **2012**, *46*, 648–652. [CrossRef]
19. Huey, R.B.; Dunham, A.E.; Overall, K.L.; Newman, R.A. Variation in locomotor performance in demographically known populations of the lizard *Sceloporus merriami*. *Physiol. Zool.* **1990**, *63*, 845–872. [CrossRef]
20. McElroy, E.J.; Bergmann, P.J. Tail autotomy, tail size, and locomotor performance in lizards. *Physiol. Biochem. Zool.* **2013**, *86*, 669–679. [CrossRef]
21. Daniels, C.B. Running: An escape strategy enhanced by autotomy. *Herpetologica* **1983**, *39*, 162–165.
22. Naidenov, L.A.; Allen, W.L. Tail autotomy works as a pre-capture defense by deflecting attacks. *Ecol. Evol.* **2021**, *11*, 3058–3064. [CrossRef] [PubMed]
23. Vitt, L.J.; Congdon, J.D.; Dickson, N.A. Adaptive strategies and energetics of tail autotomy in lizards. *Ecology* **1977**, *58*, 326–337. [CrossRef]
24. Rodrigues, M. A new species of lizard, genus *Micrablepharus* (Squamata: Gymnophthalmidae), from Brazil. *Herpetologica* **1996**, *52*, 535–541.
25. Colli, G.R.; Bastos, R.P.; Araujo, A.F.B. The character and dynamics of the Cerrado herpetofauna. In *The Cerrados of Brazil: Ecology and Natural History of a Neotropical Savannah*; Oliveira, P.S., Marquis, R.J., Eds.; Columbia University Press: New York, NY, USA, 2002; Volume 1, pp. 223–241.
26. Santos, M.G.; Nogueira, C.; Giugliano, L.G.; Colli, G.R. Landscape evolution and phylogeography of *Micrablepharus atticolus* (Squamata, Gymnophthalmidae), an endemic lizard of the Brazilian Cerrado. *J. Biogeogr.* **2014**, *41*, 1506–1519. [CrossRef]
27. de Moura, M.R.; Dayrell, J.S.; de Avelar São-Pedro, V. Reptilia, Gymnophthalmidae, *Micrablepharus maximiliani* (Reinhardt and Lutken, 1861): Distribution extension, new state record and geographic distribution map. *Check List* **2010**, *6*, 419–426. [CrossRef]
28. Gainsbury, A.M.; Colli, G.R. Lizard assemblages from natural Cerrado enclaves in southwestern Amazonia: The role of stochastic extinctions and isolation. *Biotropica* **2003**, *35*, 503–519. [CrossRef]
29. Vieira, G.H.C.; Mesquita, D.O.; Péres, A.K., Jr.; Kitayama, K.; Colli, G.R. Lacertilia: *Micrablepharus atticolus* (NCN). Natural history. *Herpetol. Rev.* **2000**, *31*, 241–242.
30. Vitt, L.J. An introduction to the ecology of Cerrado lizards. *J. Herpetol.* **1991**, *25*, 79–90. [CrossRef]
31. Vitt, L.J.; Caldwell, J.P. Ecological observations on Cerrado lizards in Rondônia, Brazil. *J. Herpetol.* **1993**, *27*, 46–52. [CrossRef]
32. de Sousa, H.C.; Soares, A.H.S.B.; Costa, B.M.; Pantoja, D.L.; Caetano, G.H.; de Queiroz, T.A.; Colli, G.R. Fire regimes and the demography of the lizard *Micrablepharus atticolus* (Squamata, Gymnophthalmidae) in a biodiversity hotspot. *S. Am. J. Herpetol.* **2015**, *10*, 143–156. [CrossRef]
33. Dal Vechio, F.; Recoder, R.; Zaher, H.; Rodrigues, M.T. Natural history of *Micrablepharus maximiliani* (Squamata: Gymnophthalmidae) in a Cerrado region of northeastern Brazil. *Zoologia* **2014**, *31*, 114–118. [CrossRef]
34. Roscito, J.G.; Nunes, P.M.S.; Rodrigues, M.T. Digit evolution in gymnophthalmid lizards. *Int. J. Dev. Biol.* **2014**, *58*, 895–908. [CrossRef]
35. Grizante, M.B.; Brandt, R.; Kohlsdorf, T. Evolution of body elongation in gymnophthalmid lizards: Relationships with climate. *PLoS ONE* **2012**, *7*, e49772. [CrossRef]
36. Renous, S.; Hofling, E.; Gasc, J.P. Respective role of the axial and appendicular systems in relation to the transition to limblessness. *Acta Biotheor.* **1998**, *46*, 141–156. [CrossRef] [PubMed]
37. Renous, S.; Hofling, E.; Gasc, J.P. On the rhythmical coupling of the axial and appendicular systems in small terrestrial lizards (Sauria: Gymnophthalmidae). *Zool.-Anal. Complex Syst.* **1999**, *102*, 31–49.
38. Höfling, E.; Renous, S. High frequency of pauses during intermittent locomotion of small South American gymnophthalmid lizards (Squamata, Gymnophthalmidae). *Phyllomedusa* **2004**, *3*, 83–94. [CrossRef]
39. Sousa, H.C.; Costa, B.M.; Morais, C.J.S.; Pantoja, D.L.; de Queiroz, T.A.; Vieira, C.R.; Colli, G.R. Blue tales of a blue-tailed lizard: Ecological correlates of tail autotomy in *Micrablepharus atticolus* (Squamata, Gymnophthalmidae) in a Neotropical savannah. *J. Zool.* **2016**, *299*, 202–212. [CrossRef]
40. Gillis, G.; Higham, T.E. Consequences of lost endings: Caudal autotomy as a lens for focusing attention on tail function during locomotion. *J. Exp. Biol.* **2016**, *219*, 2416–2422. [CrossRef] [PubMed]
41. Christian, K.A.; Tracy, C.R. The effect of the thermal environment on the ability of hatchling Galapagos land iguanas to avoid predation during dispersal. *Oecologia* **1981**, *49*, 218–223. [CrossRef]
42. Jayne, B.C.; Bennett, A.F. Selection on locomotor performance capacity in a natural population of garter snakes. *Evolution* **1990**, *44*, 1204–1229. [CrossRef]
43. Greenwald, O.E. Thermal dependence of striking and prey capture by gopher snakes. *Copeia* **1974**, *1974*, 141–148. [CrossRef]
44. Webb, W.P. Body form, locomotion and foraging in aquatic vertebrates. *Am. Zool.* **1984**, *24*, 107–120. [CrossRef]

45. Garland, T.; Hankins, E.; Huey, R.B. Locomotor capacity and social dominance in male lizards. *Funct. Ecol.* **1990**, *4*, 243–250. [CrossRef]
46. Chapple, D.G.; Swain, R. Effect of caudal autotomy on locomotor performance in a viviparous skink, *Niveoscincus metallicus*. *Funct. Ecol.* **2002**, *16*, 817–825. [CrossRef]
47. Dayananda, B.; Ibarguengoytia, N.; Whiting, M.J.; Webb, J.K. Effects of pregnancy on body temperature and locomotor performance of velvet geckos. *J. Therm. Biol.* **2017**, *65*, 64–68. [CrossRef]
48. Cloyed, C.S.; Grady, J.M.; Savage, V.M.; Uyeda, J.C.; Dell, A.I. The allometry of locomotion. *Ecology* **2021**, *102*, e03369. [CrossRef]
49. Pironon, S.; Papuga, G.; Villellas, J.; Angert, A.L.; Garcia, M.B.; Thompson, J.D. Geographic variation in genetic and demographic performance: New insights from an old biogeographical paradigm. *Biol. Rev. Camb. Philos. Soc.* **2017**, *92*, 1877–1909. [CrossRef]
50. Lin, Z.H.; Ji, X. Partial tail loss has no severe effects on energy stores and locomotor performance in a lacertid lizard, *Takydromus septentrionalis*. *J. Comp. Physiol. B-Biochem. Syst. Environ. Physiol.* **2005**, *175*, 567–573. [CrossRef]
51. Alvares, C.A.; Stape, J.L.; Sentelhas, P.C.; De Moraes Goncalves, J.L.; Sparovek, G. Köppen's climate classification map for Brazil. *Meteorol. Z.* **2013**, *22*, 711–728. [CrossRef]
52. Angilletta, M.J., Jr. *Thermal Adaptation: A Theoretical and Empirical Synthesis*; Oxford University Press: Oxford, UK, 2009; pp. 1–302. [CrossRef]
53. Wood, S.N. *Generalized Additive Models: An Introduction with R*, 2nd ed.; CRC Press, Taylor & Francis Group: Boca Raton, FL, USA, 2017; p. 476.
54. Wood, S.N. Fast stable restricted maximum likelihood and marginal likelihood estimation of semiparametric generalized linear models. *J. R. Stat. Soc. Ser. B-Stat. Methodol.* **2011**, *73*, 3–36. [CrossRef]
55. R Core Team. *R: A Language and Environment for Statistical Computing*; R Foundation for Statistical Computing: Vienna, Austria, 2019.
56. Wood, S.N. *Generalized Additive Models: An Introduction with R*; CRC Press: Boca Raton, FL, USA, 2006; Volume 16, p. 391.
57. Bartón, K. *MuMIn: Multi-Model Inference*, R package version 1.40.4; CRC Press: Boca Raton, FL, USA, 2018.
58. Leeb, H.; Potscher, B.M. Model selection and inference: Facts and fiction. *Econom. Theory* **2005**, *21*, 21–59. [CrossRef]
59. Hjort, N.L.; Claeskens, G. Frequentist model average estimators. *J. Am. Stat. Assoc.* **2003**, *98*, 879–899. [CrossRef]
60. Burnham, K.P.; Anderson, D.R. *Model Selection and Multi-Model Inference*, 2nd ed.; Springer: New York, NY, USA, 2002; p. 496.
61. Liang, H.; Zou, G.H.; Wan, A.T.K.; Zhang, X.Y. Optimal weight choice for frequentist model average estimators. *J. Am. Stat. Assoc.* **2011**, *106*, 1053–1066. [CrossRef]
62. Galipaud, M.; Gillingham, M.A.F.; David, M.; Dechaume-Moncharmont, F.-X. Ecologists overestimate the importance of predictor variables in model averaging: A plea for cautious interpretations. *Methods Ecol. Evol.* **2014**, *5*, 983–991. [CrossRef]
63. Galipaud, M.; Gillingham, M.A.F.; Dechaume-Moncharmont, F.-X. A farewell to the sum of Akaike weights: The benefits of alternative metrics for variable importance estimations in model selection. *Methods Ecol. Evol.* **2017**, *8*, 1668–1678. [CrossRef]
64. Lukacs, P.M.; Burnham, K.P.; Anderson, D.R. Model selection bias and Freedman's paradox. *Ann. Inst. Stat. Math.* **2010**, *62*, 117–125. [CrossRef]
65. Bauwens, D.; Garland, T.; Castilla, A.M.; Vandamme, R. Evolution of sprint speed in lacertid lizards: Morphological, physiological, and behavioral covariation. *Evolution* **1995**, *49*, 848–863.
66. Sinclair, B.J.; Marshall, K.E.; Sewell, M.A.; Levesque, D.L.; Willett, C.S.; Slotsbo, S.; Dong, Y.W.; Harley, C.D.G.; Marshall, D.J.; Helmuth, B.S.; et al. Can we predict ectotherm responses to climate change using thermal performance curves and body temperatures? *Ecol. Lett.* **2016**, *19*, 1372–1385. [CrossRef]
67. Angilletta, M.J., Jr.; Bennett, A.F.; Guderley, H.; Navas, C.A.; Seebacher, F.; Wilson, R.S. Coadaptation: A unifying principle in evolutionary thermal biology. *Physiol. Biochem. Zool.* **2006**, *79*, 282–294. [CrossRef]
68. Higham, T.E.; Russell, A.P.; Zani, P.A. Integrative biology of tail autotomy in lizards. *Physiol. Biochem. Zool.* **2013**, *86*, 603–610. [CrossRef] [PubMed]
69. Huey, R.B.; Stevenson, R.D. Integrating thermal physiology and ecology of ectotherms: A discussion of approaches. *Am. Zool.* **1979**, *19*, 357–366. [CrossRef]
70. Huey, R.B. Temperature, physiology, and the ecology of reptiles. In *Biology of the Reptilia*; Gans, C., Pough, F.H., Eds.; Academic Press: London, UK, 1982; Volume 12, pp. 25–91.
71. Angilletta, M.J., Jr.; Niewiarowski, P.H.; Navas, C.A. The evolution of thermal physiology in ectotherms. *J. Therm. Biol.* **2002**, *27*, 249–268. [CrossRef]
72. Strangas, M.L.; Navas, C.A.; Rodrigues, M.T.; Carnaval, A.C. Thermophysiology, microclimates, and species distributions of lizards in the mountains of the Brazilian Atlantic Forest. *Ecography* **2019**, *42*, 354–364. [CrossRef]
73. Colli, G.R.; Hoogmoed, M.S.; Cannatella, D.C.; Cassimiro, J.; Gomes, J.O.; Ghellere, J.M.; Nunes, P.M.S.; Pellegrino, K.C.M.; Salerno, P.; De Souza, S.M.; et al. Description and phylogenetic relationships of a new genus and two new species of lizards from Brazilian Amazonia, with nomenclatural comments on the taxonomy of Gymnophthalmidae (Reptilia: Squamata). *Zootaxa* **2015**, *4000*, 401–427. [CrossRef]
74. Clusella-Trullas, S.; Chown, S.L. Lizard thermal trait variation at multiple scales: A review. *J. Comp. Physiol. B-Biochem. Syst. Environ. Physiol.* **2014**, *184*, 5–21. [CrossRef]
75. Huey, R.B.; Kingsolver, J.G. Evolution of thermal sensitivity of ectotherm performance. *Trends Ecol. Evol.* **1989**, *4*, 131–135. [CrossRef]

76. Van Damme, R.; Vanhooydonck, B. Origins of interspecific variation in lizard sprint capacity. *Funct. Ecol.* **2001**, *15*, 186–202. [CrossRef]
77. Pontes-da-Silva, E.; Magnusson, W.E.; Sinervo, B.; Caetano, G.H.; Miles, D.B.; Colli, G.R.; Diele-Viegas, L.M.; Fenker, J.; Santos, J.C.; Werneck, F.P. Extinction risks forced by climatic change and intraspecific variation in the thermal physiology of a tropical lizard. *J. Therm. Biol.* **2018**, *73*, 50–60. [CrossRef]
78. Mendez-Galeano, M.A.; Paternina-Cruz, R.F.; Calderon-Espinosa, M.L. The highest kingdom of Anolis: Thermal biology of the Andean lizard *Anolis heterodermus* (Squamata: Dactyloidae) over an elevational gradient in the Eastern Cordillera of Colombia. *J. Therm. Biol.* **2020**, *89*, 102498. [CrossRef]
79. Marques, E.Q.; Marimon-Junior, B.H.; Marimon, B.S.; Matricardi, E.A.T.; Mews, H.A.; Colli, G.R. Redefining the Cerrado-Amazonia transition: Implications for conservation. *Biodivers. Conserv.* **2020**, *29*, 1501–1517. [CrossRef]
80. Gaston, K.J. Geographic range limits: Achieving synthesis. *Proc. R. Soc. B-Biol. Sci.* **2009**, *276*, 1395–1406. [CrossRef]
81. Garner, T.W.J.; Pearman, P.B.; Angelone, S. Genetic diversity across a vertebrate species' range: A test of the central-peripheral hypothesis. *Mol. Ecol.* **2004**, *13*, 1047–1053. [CrossRef]
82. Pironon, S.; Villellas, J.; Morris, W.F.; Doak, D.F.; Garcia, M.B. Do geographic, climatic or historical ranges differentiate the performance of central versus peripheral populations? *Glob. Ecol. Biogeogr.* **2015**, *24*, 611–620. [CrossRef]
83. Valladares, F.; Matesanz, S.; Guilhaumon, F.; Araujo, M.B.; Balaguer, L.; Benito-Garzon, M.; Cornwell, W.; Gianoli, E.; van Kleunen, M.; Naya, D.E.; et al. The effects of phenotypic plasticity and local adaptation on forecasts of species range shifts under climate change. *Ecol. Lett.* **2014**, *17*, 1351–1364. [CrossRef]
84. Shine, R. Effects of pregnancy on locomotor performance: An experimental study on lizards. *Oecologia* **2003**, *136*, 450–456. [CrossRef]
85. Olsson, M.; Shine, R.; Bak-Olsson, E. Locomotor impairment of gravid lizards: Is the burden physical or physiological? *J. Evol. Biol.* **2000**, *13*, 263–268. [CrossRef]
86. Bauwens, D.; Thoen, C. Escape tactics and vulnerability to predation associated with reproduction in the lizard *Lacerta vivipara*. *J. Anim. Ecol.* **1981**, *50*, 733–743. [CrossRef]
87. Itonaga, K.; Jones, S.M.; Wapstra, E. Do gravid females become selfish? Female allocation of energy during gestation. *Physiol. Biochem. Zool.* **2012**, *85*, 231–242. [CrossRef]
88. Sinervo, B.; Hedges, R.; Adolph, S.C. Decreased sprint speed as a cost of reproduction in the lizard *Sceloporus Occidentalis*: Variation among populations. *J. Exp. Biol.* **1991**, *155*, 323–336. [CrossRef]
89. Seigel, R.A.; Huggins, M.M.; Ford, N.B. Reduction in locomotor ability as a cost of reproduction in gravid snakes. *Oecologia* **1987**, *73*, 481–485. [CrossRef]
90. Shine, R. "Costs" of reproduction in reptiles. *Oecologia* **1980**, *46*, 92–100. [CrossRef] [PubMed]
91. Van Damme, R.; Bauwens, D.; Verheyen, R.F. Effect of relative clutch mass on sprint speed in the lizard *Lacerta vivipara*. *J. Herpetol.* **1989**, *23*, 459–461. [CrossRef]
92. Cooper, W.E., Jr.; Wilson, D.S.; Smith, G.R. Sex, reproductive status, and cost of tail autotomy via decreased running speed in lizards. *Ethology* **2009**, *115*, 7–13. [CrossRef]
93. Shine, R. Locomotor speeds of gravid lizards: Placing 'costs of reproduction' within an ecological context. *Funct. Ecol.* **2003**, *17*, 526–533. [CrossRef]
94. Clemente, C.J.; Withers, P.C.; Thompson, G. Optimal body size with respect to maximal speed for the yellow-spotted monitor lizard (*Varanus panoptes*; Varanidae). *Physiol. Biochem. Zool.* **2012**, *85*, 265–273. [CrossRef]
95. Jones, J.H.; Lindstedt, S.L. Limits to maximal performance. *Annu. Rev. Physiol.* **1993**, *55*, 547–569. [CrossRef]
96. Heglund, N.C.; Taylor, C.R.; McMahon, T.A. Scaling stride frequency and gait to animal size: Mice to horses. *Science* **1974**, *186*, 1112–1113. [CrossRef] [PubMed]
97. Schmidt-Nielsen, K. Locomotion: Energy cost of swimming, flying, and running. *Science* **1972**, *177*, 222. [CrossRef]
98. Zamora-Camacho, F.J.; Reguera, S.; Moreno-Rueda, G. Does tail autotomy affect thermoregulation in an accurately thermoregulating lizard? Lessons from a 2200-m elevational gradient. *J. Zool.* **2015**, *297*, 204–210. [CrossRef]
99. Downes, S.J.; Shine, R. Why does tail loss increase a lizard's later vulnerability to snake predators? *Ecology* **2001**, *82*, 1293–1303. [CrossRef]
100. Jagnandan, K.; Russell, A.P.; Higham, T.E. Tail autotomy and subsequent regeneration alter the mechanics of locomotion in lizards. *J. Exp. Biol.* **2014**, *217*, 3891–3897. [CrossRef] [PubMed]
101. Hamley, T. Functions of the tail in bipedal locomotion of lizards dinosaurs and pterosaurs. *Mem. Qld. Mus.* **1990**, *28*, 153–158.
102. Höfling, E.; Renous, S.; Curcio, F.F.; Eterovic, A.; de Souza Santos Filho, P. Effects of surface roughness on the locomotion of a long-tailed lizard, *Colobodactylus taunayi* Amaral, 1933 (Gymnophthalmidae: Heterodactylini). *Int. J. Zool.* **2012**, *2012*, 1–16. [CrossRef]
103. Vitt, L.J.; Caldwell, J.P. *Herpetology. An Introductory Biology of Amphibians and Reptiles*, 4th ed.; Elsevier, Inc.: London, UK, 2014.
104. Gans, C. Tetrapod limblessness: Evolution and functional corollaries. *Am. Zool.* **1975**, *15*, 455–467. [CrossRef]
105. da Silva, F.O.; Fabre, A.C.; Savriama, Y.; Ollonen, J.; Mahlow, K.; Herrel, A.; Muller, J.; Di-Poi, N. The ecological origins of snakes as revealed by skull evolution. *Nat. Commun.* **2018**, *9*, 11. [CrossRef]
106. McElroy, E.J. The effect of tail autotomy on locomotor performance in the long tailed grass lizard, Takydromus sexlineatus. *Integr. Comp. Biol.* **2011**, *51*, E89.

107. Wiens, J.J.; Slingluff, J.L. How lizards turn into snakes: A phylogenetic analysis of body-form evolution in anguid lizards. *Evolution* **2001**, *55*, 2303–2318. [CrossRef]
108. Wiens, J.J.; Brandley, M.C.; Reeder, T.W. Why does a trait evolve multiple times within a clade? Repeated evolution of snakelike body form in squamate reptiles. *Evolution* **2006**, *60*, 123–141. [CrossRef]

MDPI
St. Alban-Anlage 66
4052 Basel
Switzerland
Tel. +41 61 683 77 34
Fax +41 61 302 89 18
www.mdpi.com

Diversity Editorial Office
E-mail: diversity@mdpi.com
www.mdpi.com/journal/diversity

www.ingramcontent.com/pod-product-compliance
Lightning Source LLC
LaVergne TN
LVHW070611170726
843515LV00004B/46

* 9 7 8 3 0 3 6 5 4 0 5 1 1 *